INTERNATIONAL ENERGY AGENCY

FLEXIBILITY IN NATURAL GAS SUPPLY AND DEMAND

INTERNATIONAL ENERGY AGENCY
9, rue de la Fédération,
75739 Paris, cedex 15, France

ORGANISATION FOR
ECONOMIC CO-OPERATION
AND DEVELOPMENT

The International Energy Agency (IEA) is an autonomous body which was established in November 1974 within the framework of the Organisation for Economic Co-operation and Development (OECD) to implement an inter-national energy programme.

It carries out a comprehensive programme of energy co-operation among twenty-six* of the OECD's thirty Member countries. The basic aims of the IEA are:

- to maintain and improve systems for coping with oil supply disruptions;

- to promote rational energy policies in a global context through co-operative relations with non-member countries, industry and international organisations;

- to operate a permanent information system on the international oil market;

- to improve the world's energy supply and demand structure by developing alternative energy sources and increasing the efficiency of energy use;

- to assist in the integration of environmental and energy policies.

IEA Member countries: Australia, Austria, Belgium, Canada, the Czech Republic, Denmark, Finland, France, Germany, Greece, Hungary, Ireland, Italy, Japan, the Republic of Korea, Luxembourg, the Netherlands, New Zealand, Norway, Portugal, Spain, Sweden, Switzerland, Turkey, the United Kingdom, the United States. The European Commission also takes part in the work of the IEA.

Pursuant to Article 1 of the Convention signed in Paris on 14th December 1960, and which came into force on 30th September 1961, the Organisation for Economic Co-operation and Development (OECD) shall promote policies designed:

- to achieve the highest sustainable economic growth and employment and a rising standard of living in Member countries, while maintaining financial stability, and thus to contribute to the development of the world economy;

- to contribute to sound economic expansion in Member as well as non-member countries in the process of economic development; and

- to contribute to the expansion of world trade on a multilateral, non-discriminatory basis in accordance with international obligations.

The original Member countries of the OECD are Austria, Belgium, Canada, Denmark, France, Germany, Greece, Iceland, Ireland, Italy, Luxembourg, the Netherlands, Norway, Portugal, Spain, Sweden, Switzerland, Turkey, the United Kingdom and the United States. The following countries became Members subsequently through accession at the dates indicated hereafter: Japan (28th April 1964), Finland (28th January 1969), Australia (7th June 1971), New Zealand (29th May 1973), Mexico (18th May 1994), the Czech Republic (21st December 1995), Hungary (7th May 1996), Poland (22nd November 1996), the Republic of Korea (12th December 1996) and Slovakia (28th September 2000). The Commission of the European Communities takes part in the work of the OECD (Article 13 of the OECD Convention).

FOREWORD

Flexibility of natural gas supply is essential for efficient operation of the gas industry. Gas companies have developed various flexibility tools to harmonise the variations and fluctuations in gas demand with the relative rigidity of gas supply. These tools mainly work on the volumes of supply and demand.

Market liberalisation is changing the traditional landscape. It brings a new market-driven approach to flexibility. In competitive markets, new market mechanisms are available to bring gas supply and demand into line, for example in marketplaces where both gas and flexibility services are traded. However, inelastic demand by non-interruptible customers still needs to be covered as it arises. Traditional flexibility tools remain important for this purpose.

This book surveys the impact of market opening on flexibility tools and mechanisms. The experience in competitive markets demonstrates the changes which are taking place. It is important to keep track of these developments in newly opened markets and this book suggests some developments that governments should particularly monitor: the parallel opening to competition of the electricity and gas sectors and the possible impact on gas supply and demand of the arbitrage between gas and power; the increased volatility of prices; and the responsibility of the different players.

This book is one part of a more general review by the International Energy Agency of the security of natural gas supply in OECD countries.

This book is published under my authority as Executive Director of the International Energy Agency.

Robert Priddle
Executive Director

ACKNOWLEDGEMENTS

The main author of this book is Sylvie Cornot-Gandolphe. Ralf Dickel, Head of the Energy Diversification Division, directed the project and provided valuable input and advice. The Director of the Long-Term Office, Olivier Appert, provided support and encouragement throughout the project. Special thanks to all other IEA colleagues who provided comments on the analysis, statistical data and editorial oversight.

Comments and suggestions from various delegates to the Standing Group on Long-term Co-operation as well as from key experts of the gas industry are gratefully acknowledged.

TABLE OF CONTENTS

LIST OF TABLES IN TEXT

LIST OF FIGURES IN TEXT

LIST OF BOXES IN TEXT

EXECUTIVE SUMMARY

GAS DEMAND AND SUPPLY DO NOT ALWAYS REACT TO MARKET SIGNALS IN WAYS TYPICAL OF OTHER COMMODITIES

Gas demand reacts only partially to price signals

In commodity markets, supply and demand are normally balanced by the price mechanism. Buyers and sellers react to price signals given by the market. The need for flexibility of volume – the ability to add to supply or reduce demand – reflects the nature of the commodity. In gas markets, demand is not particularly flexible. Most residential and commercial customers are unable to switch easily to alternative fuels. Once gas consumers have committed to using gas as a fuel and have invested in gas-fired equipment, they cannot then change quickly to other fuels. Furthermore, these customers cannot store gas. They therefore cannot react easily to market signals such as a sudden price increase. In other words, these customers have a rather price-inelastic demand.

On the other hand, industrial gas customers equipped with bi-fuel equipment or power generators linked to the power grid can replace gas on a short-term basis, in response to price signals. However, in markets where the price of gas is linked to the price of replacement fuels, these customers actually have little incentive to switch to another fuel. This is indeed the case for many European industrial customers, who as a result have little incentive to maintain costly bivalent equipment. Where industrial consumers do hold stocks of alternate fuels, they usually hold only enough fuel to cover short periods of gas interruption, thus limiting the scope and effect of fuel-switching that may take place.

Gas supply does not always react to price signals either

In most parts of the world, gas production and transportation require large investments because of difficult geological conditions of extraction and production and the increasing distances over which gas must be transported from wellhead to market. In addition, once a gas project is planned out and the necessary investment funds are committed, the project's carrying capacity is usually fixed. This is particularly so because gas infrastructure takes a considerable time to construct and bring on line, and therefore additions are generally made in major increments over time, as individual projects are initiated, constructed and completed. In traditional markets, major investments of this sort are most often handled by large utilities. It is important for utilities to conclude additional long-term delivery and transportation contracts with gas suppliers and transporters in order to keep up with increasing demand and to retain sufficient flexibility to adapt to fluctuations in that demand.

In liberalised markets, supply reacts to price signals sent by the market. This means an optimum use of spare capacity. Higher prices will also trigger investments into additional production capacity, as suggested above. But even in liberalised gas markets, flexibility of supply still has distinct limits. In the United States, for instance, when gas prices are high, producers increase their drilling. However, it takes about 18 months before such new drilling translates into additional production capacity. The time lag can be even longer for large, capital-intensive projects, such as LNG terminals and long-distance pipelines. Moreover, when more than one country is involved in the gas chain, price signals given by the market in a consuming country will not always translate into new investment in exploration, production and transport capacity in supplying countries.

Flexibility is an essential component of supply security

Residential and commercial gas demand is seasonal, temperature-dependent and inelastic. It must be covered as it arises. To respond to such variations requires flexibility. Flexibility here means the ability to adapt supply to foreseeable volume variations in demand (mainly seasonal) and to adjust for erratic fluctuations in demand (mainly short-term temperature variations), or to adapt demand (i.e., reduce it) when supply is insufficient. Flexibility is achieved through physical instruments and contractual arrangements that anticipate likely variations in demand and balance the volume of gas supply and demand at any time. Physical instruments include variable supply in production and import contracts, gas from storage and line-pack. Contractual arrangements can take the form of contracts with "interruptible customers" that allow interruption of their supply at agreed times. Liberalised markets bring a new market-driven approach to flexibility that supplements the traditional flexibility tools.

The notion of flexibility is usually linked both to volumes and to delivery or transmission capacities. The latter are becoming more critical because the unbundling of the supply and transmission businesses creates a new risk – that a transmission network's capacity will not always be sufficient to carry all the available supply. Thus, flexibility to produce a certain amount of additional supply is not in itself sufficient to meet unexpected requirements. Enough extra capacity must be available on the transmission grid to transport this increment in a timely way. Recently, balancing supply and demand by market-based mechanisms has become an important feature. Throughout this survey, however, the term flexibility refers to flexibility in volumes and capacities aimed at balancing supply and demand.

HOW FLEXIBILITY REQUIREMENTS ARE EVOLVING TO MEET SUPPLY AND DEMAND TRENDS IN THE OECD

Increasing gas demand in all three OECD regions requires additional investment to meet greater fluctuation in demand. The ability to respond varies from region to region

OECD gas demand is expected to increase rapidly during the next 30 years. The driving force will be increased use of gas in power generation, mainly in combined-cycle gas turbine power plants. Residential, commercial and combined heat and power sectors are also expected to increase their gas consumption. Because demand from these sectors is highly sensitive to temperatures, fluctuations in gas demand will be larger in the three OECD regions as demand increases.

On the other hand, supply to the three regions will remain fairly rigid. World gas reserves are concentrated in only a few regions and countries, mainly outside the IEA. As a result, most IEA countries will become increasingly dependent on imports from more distant sources. Further diversification of supply sources and the increased use of flexibility instruments will be required to avoid vulnerability to specific supply sources and routes as well as undue exposure to specific supply risks.

An important issue is whether adequate and timely investment in upstream production and transportation capacity will be triggered by market signals as supply gaps develop. The answer differs fundamentally from one OECD region to another. For decades, the reserves-to-production ratio in the US has been just 7-8 years, as new reserves were developed "just in time". Accelerating depletion rates may make this more challenging. In the recent past, circumstances in the UK resembled those in the US, but this may change, as resources on the UK Continental Shelf are now on the decline. Continental Europe and Japan have long relied on long-haul imports of pipeline gas and LNG, with long-term contracts that, in effect, underpin the capital investment in supplier countries mainly outside the OECD.

In OECD North America, gas-to-gas competition has developed. Supply and demand are matched by market mechanisms, and several different flexibility tools compete on the market. The market is liquid and self-sufficient, and customers can always buy flexibility on the market, although prices for these services can be high.

The UK is in a similar position, with a fully liberalised, competitive and currently self-sufficient gas market. So far, UK gas supply has reacted to price signals sent by the market. With the depletion of gas reserves, however, the country's ability to meet its own needs for flexibility is likely to decline.

In Continental Europe, a high level of flexibility is in place. However, the capability to cope with any substantial supply disruption has never had to be tested. For now, gas oversupply and some spare capacity in the gas transmission grid allow a great deal of flexibility of supply, although the situation varies widely among European countries. There are some bottlenecks on the EU transportation network, and some countries rely on just one source of supply and so enjoy little supply flexibility.

Many Continental European countries have ample storage capacity, enough to supply the market for several months. Here again, though, there are large differences among countries. Italy, for instance, has a highly adaptable system that includes a large amount of strategic storage to deal with supply interruptions. Other countries, however, use storage only for balancing and some countries have little or no storage capacity at all.

The introduction of competition, with third-party access[1] in the gas sector, has fostered market mechanisms to match supply and demand. However, because of Europe's large and increasing import dependency on long-haul gas, traditional flexibility instruments (such as flexibility in production and import contracts, storage and interruptible contracts) will remain vital. Increasing demand will gradually reduce spare capacity in transportation and supply. More storage capacity and more interconnection among the gas grids thus will be required. The new ways in which the liberalised market will assist countries in meeting

---

1 Third-party access (TPA) is the right or opportunity for a third party (shipper) to make use of the transportation or distribution services of a pipeline company to move gas for a set or negotiated charge.

their flexibility requirements will effectively supplement the extended use of traditional instruments of flexibility.

Improved interconnection of the gas transmission network will be an important factor in enhancing flexibility and security. It can allow additional gas to be transported in the short term from one market to a nearby one – for example, when extreme weather in one market coincides with mild weather in another.

In the OECD Pacific region, the two LNG-importing countries, Japan and Korea, have very different seasonal profiles. Japan, which uses large amounts of gas to generate electricity but does not rely directly on it for space heating, is less subject to seasonal variations than is Korea, which uses less gas overall but does use it mainly for space heating. Both countries have developed several tools to ensure flexibility and security of supply in their gas systems. Japan relies on a combination of several means, such as modular supply and delivery systems that limit dependence on any single installation, along with opportunities for fuel substitution and sharing via the electricity generation system. Korea relies heavily on buying spot LNG cargoes to cover its peak winter demand. In the future, seasonal variations are likely to increase as a result of increased use of gas in the residential and commercial sectors in both countries, and for upper and middle load production in the power generation sector in Japan. This trend will have to be met by increased flexibility of supply.

Flexibility requirements continue to differ between the three regions, and their use of flexibility instruments varies commensurately

Flexibility requirements and mechanisms differ greatly from one OECD region to another, and from one country to another, depending on the state of market liberalisation, the demand structure, the share of gas in the energy mix, the existence and size of national gas resources, and the diversification of supply. Moreover, flexibility needs and mechanisms are not static. They evolve with time. For instance, flexibility in supply declines when a major gas field, as it is depleted, can no longer offer above-average monthly deliveries, or "swing" (here defined as the maximum gas monthly delivery divided by the average monthly delivery in a given year).

The variations in flexibility provisions at present are illustrated by the case studies of IEA countries presented in annexes. They can be categorised as follows:

- producing countries with short-haul transportation, such as the Netherlands, where flexibility in demand is met by variations in production;

- producing countries with a long-distance transportation grid, such as the United States, where flexibility in demand is provided by storage and interruptible contracts;

- importing countries that meet flexibility requirements by flexibility in importing contracts, as in Belgium;

- importing countries that meet flexibility requirements mostly by storage, as in France;

- importing countries that meet flexibility requirements by a combination of flexibility in production, import contracts and storage, such as Germany; and

- LNG-importing countries with high seasonal variations in demand, such as Korea, that achieve flexibility through medium-term contracts and spot LNG cargoes.

TRADITIONAL FLEXIBILITY TOOLS

A variety of tools is available to balance supply and demand for gas at any time. These tools fall into three categories: those aimed at increasing supply flexibility, those that provide buffer stocks and those that reduce demand at times of peak gas use.

The supply side

On the supply side, the obvious response to a sudden need for additional gas is to increase one's own gas production, if possible, and to increase the volume of gas received under domestic or import purchase contracts, if possible, within any entitlement not fully used. In Europe, the Dutch gas field, Groningen, has played the role of the swing supplier, thanks both to its geological characteristics and to its proximity to the main European markets.

Imports by pipeline from remote areas offer very limited swing, because of the high share of fixed transportation costs. This is typically the case for gas received in Europe from outside the EU or for Canadian gas imported into the US. Although LNG supplies in the Pacific area were very rigid in the early days of LNG trade, they have recently become more flexible thanks mainly to the existence of spare capacity at new liquefaction plants.

In many gas systems, hourly variations in demand can often be balanced out by line-pack (i.e., by raising the pressure within a gas pipeline system above the required delivery pressure in order to increase the system's storage capability). The role that line-pack plays in balancing differs significantly among countries according to the differing designs of their transmission grids and to their particular supply patterns. For instance, in the United Kingdom, line-pack has traditionally been able to meet incremental demand of up to 3% of total demand, but in Spain the figure is just 0.4%.

Storage facilities also play an important role in providing flexibility to meet seasonal variations, to cover unforeseen changes in demand and in some countries as a back-up in case of supply interruptions. In traditional markets, withdrawals from storage facilities have been the major instrument used to provide supplementary gas supplies when the country's own production or contract flexibility are insufficient to bridge the gap. LNG receiving facilities also offer supply flexibility and storage possibilities for LNG-importing countries beyond the flexibility offered by the import contracts.

OECD North America and Europe have ample storage capacity. North America has 453 underground gas storage facilities, representing 17% of annual gas consumption; OECD Europe has 94, representing 13%, but distribution of storage is uneven across the market. Storage helps to balance rigid gas supplies and seasonal demand while minimising the size of the pipelines needed and the transportation costs. Some European companies also maintain strategic reserves beyond their commercial requirements for security-of-supply reasons. In North America, storage is increasingly used for trading purposes.

The demand side

On the demand side, gas suppliers often conclude interruptible contracts with large industrial consumers and power generators. In return for a discount on the price they pay for gas, these customers take on a contractual obligation to stop taking gas under specified conditions usually linked to outside temperatures. Interruptible customers may install dual-fired capacity so that they can switch from gas to alternative fuels in times of interruption, or – in the case of gas-fired power generation – they may arrange for back-up supplies of electricity from the grid. Alternatively, of course, they may simply stop their gas-based operation in response to a contractual interruption.

Not all these tools are available or used in every IEA country. The annexes to this study detail seasonal gas demand in each IEA country and the way it is balanced by traditional flexibility tools.

At present, supply flexibility, volumes of storage and the number of interruptible contracts are high in the IEA countries – although the situation differs for individual countries. The current situation generally provides a sufficient basis for flexibility of supply in relation to overall demand, although there remain questions about the practical limit of interruptible contracts.

THE ROLE OF FLEXIBILITY IS EVOLVING WITH MARKET LIBERALISATION

The development of trading hubs as commodity markets

With markets opening to competition, price is becoming a new instrument to balance supply and demand. Open access to infrastructure has given birth to marketplaces for gas. Trading hubs will generally emerge where several pipelines meet, often near storage sites and areas of high demand. In the UK, the entire national grid has become a virtual single market for gas. Spot markets have evolved along with these trading hubs.

In the United States and Canada, the regulatory reforms of the 1980s and 1990s fostered market hubs and centres where gas, transportation and storage rights are traded among a diverse group of market participants. At least 39 centres are now operating in North America. Some hubs have grown into full-fledged commodity markets, initially spot trading for prompt and forward delivery, and later for financial instruments such as futures and options.

In Continental Europe, third-party access prompted the development of the first trading centre, Zeebrugge in Belgium. Zeebrugge is the landing point of the UK-Belgian Interconnector and the Norwegian Zeepipe, and it also has an LNG terminal. A second hub is emerging at Bunde in Germany, close to the gas delivery point for Dutch and Norwegian gas in Germany.

More choice and responsibility for the consumer

With third-party access and unbundling of businesses, buyers may not only choose their supplier but may also choose to buy different services à la carte instead of a standard menu of services. With the unbundling of services, opportunities open up for trading not only gas but also other services, including back-up and seasonal flexibility. The Dutch company Gasunie was the first of the Continental gas companies to offer such unbundled services à la carte. Most other gas companies have followed suit. This in turn has led to the emergence of a market for flexibility services, which are offered and priced independently from the gas.

Eligible customers – gas users that meet criteria specified in the EU Gas Directive or in national legislation, such as a minimum volume of gas consumed per year, have the right to choose their supplier and request third-party access to the grid – will have to assume increasing responsibility for their flexibility choices. They may, for example, have to face the fact that interruptible contracts will increasingly be interrupted. If the market is sufficiently liquid, customers can always obtain the gas and flexibility services they want at short notice, although possibly at a higher price.

Better use of infrastructure and the effectiveness of price signals

Liberalised markets offer incentives to use infrastructure very intensively. In the United States, production, transport and storage are increasingly used at nearly full capacity. Pipeline utilisation rates in parts of the West have recently been well above 95% on a continuing basis.

Lack of supply and pipeline capacity can result in price hikes during peak periods. Sudden price hikes for gas may be a warning signal that supply capacity is not sufficient to cover the inelastic demand or unanticipated demand surges.

The new role of storage

Storage is one of several instruments providing flexibility. In liberalised gas markets, it competes with other flexibility services and instruments, such as supply flexibility, interruptible contracts, line-pack and LNG peak-shaving units. The market can now value the services provided by storage.

Storage is acquiring a new role as additional storage services appear. The traditional functions of meeting seasonal demand fluctuations, using the grid more efficiently and providing security of supply are still valid. In liberalised markets, storage will also play roles in price hedging and as a trading tool, allowing players to exploit price differentials. Governments will need to assess whether commercially motivated storage will meet the broader security needs of national energy policy.

Arbitrage between gas and power markets

The parallel opening of the electricity and gas markets provides a new market-driven method of balancing supply and demand. With the increased use of gas for power generation and the opening of the electricity grids to competition, gas demand becomes more price elastic in the short term. The short-term price elasticity of gas demand increases further as the gas grid is opened to competition.

With the development of marketplaces in both the gas and electricity sectors, arbitrage between gas and power has become possible and is increasingly used. Through arbitrage between gas and electricity markets, gas demand and supply can be balanced by cross trading for supply and demand in the electricity sector.

Electricity producers optimise their fuel input and electricity output according to the prices of both. Generators can resell gas they have bought under long-term contracts and buy electricity from the grid when doing so can provide a higher yield than if the gas were used to produce power. This practice of arbitrage is still quite new in the US and the UK. It will certainly spread to Continental Europe.

In the US and the UK, increased use of natural gas for power generation has caused gas prices and electricity prices to influence one another. Developments in the electricity market and arbitrage between gas and power greatly affect flexibility needs and the availability of gas. Yet, winter peak in electricity demand often corresponds with the peak in gas demand for heating purposes, which could lead to systemic problems. Ensuring a secure supply of electricity and gas to the final consumers now requires close monitoring of both the electricity and gas markets and the potential for interactions between them that result from arbitrage.

NEW FLEXIBILITY PATTERNS IN THE LNG MARKET

Several factors are changing LNG markets, bringing new challenges and opportunities to both buyers and sellers. Cost reductions in liquefaction plants have made it possible for investors to secure financing by selling only part of planned capacity under long-term contracts while retaining some non-committed capacity to be sold spot or short-term or long-term at a later date. The increasing number of LNG tankers – the fleet's capacity will be increased by half in the next two to three years – has led to a more fluid market in LNG transportation. Spare capacity and third-party access to receiving terminals further increase liquidity.

New ways of conducting LNG trade are emerging. Spot and short-term transactions are increasing. Swap agreements are developing and arbitrage between regional markets is taking place to capture the price differentials between markets. An additional element of flexibility is likely to appear based on the fact that seasonal patterns in LNG demand vary from one importing country to another. Parts of the LNG business are moving away from bilateral long-term contracts towards a structure that is more flexible and more responsive to market signals.

LNG buyers may now meet part of their flexibility needs directly on the LNG market. They may do so through spot transactions, for uncovered winter peak demand, or through swaps agreements with other regions or other customers.

Traditional long-term LNG contracts are gradually being complemented by LNG transactions that are more flexible in timing and location. These transactions are starting to serve as transmitters of price signals between regional gas markets. LNG spot trading, which represented 8% of global LNG trade in 2001, will develop further, but it will not replace long-term deals entirely, as these deals will underlie new project investment. In any event, chances are that a global LNG market, such as those for oil or coal, will gradually emerge as LNG trade expands and involves more players.

RECOMMENDATIONS

The opening of the gas sector has created a situation in which parts of demand and supply are balanced by the price mechanism. It also led to the development of flexibility services that are traded in a competitive environment and are therefore valued by the market. Yet, customers with no fuel-switching capabilities and whose demand is price-inelastic still need secure, uninterrupted supply. Providing that supply requires clear responsibilities and adequate flexibility instruments. In the current context, the impact on flexibility in the gas sector of sectors that are intertwined with it – mainly the electricity sector – must be considered. Governments need to (1) ensure that the new markets can work in an unimpeded and efficient way; (2) define clear roles and responsibilities for providing the flexibility necessary to meet the varying demand of customers with no price elasticity; and (3) monitor the overall operational and investment performance of the gas sector in providing such flexibility and, if necessary, take timely action using market mechanisms.

The following recommendations apply in general, but the emphasis on specific recommendations differs greatly among the three IEA regions and between countries within a region. They vary by the degree of liberalisation of gas (and electricity) markets as well as by demand structure and supply options in the particular cases.

ENSURE THE WORKING OF THE NEW MARKETS

- Governments should allow market players to develop marketplaces and flexibility services as the market players see fit. Governments should provide a clear framework and instruments of control to prevent players from abusing their market power and to achieve other objectives of government policy. They should especially monitor the risk and impact of newly evolving financial instruments.

- Governments should ensure that flexibility services are offered on a non-discriminatory basis. While the instruments that provide flexibility, such as storage or production, are not natural monopolies, there may be market structures for flexibility services that are de facto monopolies or that may be dominated by just a few players. In such cases, third-party access or obligatory divestiture may be required to develop a market for flexibility services.

- Given the strong links between the gas and power markets, governments should ensure overall consistency in the regulation of both markets.

- Governments should work to streamline the formalities of building LNG terminals within their jurisdiction, including the efficiency of environmental reviews. Governments might consider how to improve the efficiency and economies of international LNG trade by fostering more technical standardisation and a common trading framework.

ENSURE THE PROVISION OF THE FLEXIBILITY NEEDED BY NON-INTERRUPTIBLE CUSTOMERS THAT LACK FUEL-SWITCHING CAPACITY

- Inelastic demand by non-interruptible customers must be covered. Governments should define minimal criteria and clear rules for who has to provide the necessary flexibility to meet such demand, and they should specify how it should be paid for.

- Governments should see that enough flexibility is provided to cover fluctuations and variations of non-interruptible customers' consumption. Governments should ensure that market players responsible for providing such flexibility have the necessary capacity and volumes to meet the defined criteria.

- Governments should take into account the increasing convergence between the gas and electricity sectors and the possibility of the convergence of threats to the security of supply emerging in both systems. Special attention should be paid to systemic failures, especially in extreme weather conditions, and to the effects of short-term swapping of gas between the two systems through arbitrage.

MONITOR INVESTMENT PERFORMANCE AND TAKE ACTION, IF NEEDED

■ Governments should monitor the overall investment performance of the gas sector, and where they detect forthcoming shortages of flexibility, they should identify policy action that builds on market mechanisms to the extent possible. Governments should take into account the interaction of flexibility in the gas and electricity sectors and their collective impact on investment needs.

■ Gas importers should create conditions that will induce private investors to invest in gas production, export and cross-border transportation and to expand access to reserves. Measures might include long-term contracts or joint ventures to market or produce the gas.

■ Balancing supply and demand by the price mechanism may lead to volatility if supply is tight and can not respond to market signals. Governments should identify policies that will stimulate investment into the entire gas chain. Governments should seek to mitigate technical or regulatory bottlenecks that inhibit flexibility. Here, as elsewhere, they should prevent any abuse of market power.

INTRODUCTION

WHY FLEXIBILITY IS NEEDED

In commodity markets, where demand is price sensitive, supply and demand are usually balanced by the price mechanism. Buyers and sellers react to price signals from the market. Volume flexibility, the ability to add supply or withhold demand, is not normally needed to match supply and demand.

But, for many gas customers, especially households, things are very different. Once they have chosen gas-fired equipment, they cannot alter their decision in the short run. What they want is reliable, uninterrupted gas supply that varies according to their needs. Most of these customers use gas mainly for heating. Since their gas use depends directly on the weather, they have little or no control over when they use it, although they can and do choose to have more or less heating. The inconvenience and discomfort that occur when their volume requirements are not met is very high compared to the price of the gas. As a result, their demand is relatively price-inelastic. On the other hand, industrial customers equipped with bi-fuel equipment or power generators linked to the power grid, do have the possibility of replacing gas on a short-term basis. In markets where the price of gas is linked to the price of the replacement fuels, however, customers have little incentive to reduce their gas demand by switching to another fuel. Under these circumstances supply and demand are often matched by the use of flexibility instruments, which adjust volumes on the supply side and to a smaller extent on that of demand.

In liberalised gas markets in the US and UK, the price mechanism has been playing an increasing role in making supply and demand match short-term. The household sector, however, still needs volume flexibility to meet variations and fluctuations in demand.

LARGE VARIATIONS IN DEMAND; A CERTAIN RIGIDITY IN SUPPLY

Natural gas plays a growing role as an OECD energy source. OECD gas consumption reached 1,369 bcm in 2000 which represented 22% of total primary energy supply, up from 19% in 1980. Within twenty years, gas consumption has increased by almost 50%. This can be explained by the intrinsic properties of natural gas (clean combustion, easy handling, efficiency, flexibility), and by its abundant reserves (178 tcm of global proven reserves as of 1 January 2002, representing 60 years of production at current rates). Gas meets some environmental concerns, as the effects of its use on the air and climate are less than those of other fossil fuels. Gas is price competitive and its attraction has been enhanced by the development of high-performance technologies, in particular gas turbines and combined-cycle gas turbine power plants.

Unlike oil, which is primarily used in the transportation sector, gas is mainly used for stationary purposes at a fixed site, such as space heating in residential and commercial buildings, as a feedstock for the petrochemical industry, as process gas, to produce steam for industrial purposes and for power generation. Residential and commercial sectors absorb 35% of OECD gas consumption, the figure reaching 40% in OECD Europe, 34% in OECD North America and 23% in OECD Pacific. Natural gas has been increasingly used to generate electricity although the share of power generation based on gas varies largely among IEA countries from zero in Norway to 55% in the Netherlands. There are important differences in gas-consumption patterns in the three OECD regions, as illustrated in the following table.

Table 1: Breakdown of Gas Consumption by Sector (2000 data)

	OECD Total		North America		Pacific		Europe	
	bcm	%	bcm	%	bcm	%	bcm	%
Residential/ Commercial (1)	483.85	35	263.70	34	28.28	23	191.87	40
Industry, including Raw Material	347.29	25	182.24	24	24.85	20	140.19	30
Power Generation (2)	391.78	29	210.81	27	66.86	53	114.11	24
Others (3)	145.80	11	113.38	15	5.53	4	28.91	6
Total	**1,368.72**	**100**	**768.13**	**100**	**125.52**	**100**	**475.08**	**100**

(1) Including agriculture.
(2) Including combined heat and power generation.
(3) Energy sector, district heating (accounting for 7 bcm in 2000 for the OECD as a whole), transportation sector and distribution losses.
Source: IEA (2002b).

Where gas is heavily used for space heating, total gas demand varies strongly with outside temperatures. In France, for example, consumption between the annual peak-load days and those with the lowest demand can vary by an order of magnitude. The French industry's maximum daily dispatch of 2.42 TWh or 210 mcm, was recorded on 2 January 1997. On that day, storage facilities supplied 52% of the demand.

Two aspects of variability on the demand side need to be distinguished:

(1) Variations due to demand patterns, which repeat themselves at regular intervals. The leading example is seasonality, but there are also patterns induced by social habits, such as vacation periods or weekly work schedules. These resulting demand variations are to a large extent foreseeable.

(2) Variations due to exogenous forces, mainly from fluctuations in temperature, which trigger a corresponding fluctuation in the demand for heating or cooling. While it is predictable that winter days are colder than summer days, each winter day may vary in temperature in an unforeseeable way.

The flexibility required to meet repetitive demand variation is foreseeable, so that economic decisions can be taken with some certainty. But the flexibility required to meet unexpected fluctuations takes on the character of a physical-risk insurance. It requires economic decisions to be taken in the face of greater uncertainty. Residential and commercial customers rightly expect that gas will be supplied even on the coldest winter day. Back-up systems are too expensive for these customers and an interruption of gas supplies on such a day would cause unacceptable discomfort, and possibly real damage. In addition, if gas distributors were unable to guarantee supply in extreme conditions, alternative fuels such as heating oil would gain tremendous competitive advantage, which could probably not be offset by lower gas prices. Therefore, the gas industry is required to take precautions to be able to meet the coldest day of winter demand by non-interruptible customers, without knowing when or how often that case will occur. Data on past temperatures are usually available for several decades as well as models reflecting consumers' reactions to different weather conditions. But the investment required to meet such cases may only be recovered by charging a risk premium to customers requiring such security of supply.

Gas suppliers must cover a seasonally-determined and temperature-dependent demand, which is price inelastic and which must be covered as it arises. There are additional flexibility requirements on the supply side to anticipate interruption due to technical, contractual or political reasons, or even because of labour-management friction.

In this study, the term "flexibility" refers to the capability to change gas volumes over defined time-periods. Flexibility may be used to adapt supply to variations and fluctuations in demand or to adapt elements of demand in the case of insufficient supply.

In the past, the provision of flexibility did not affect the price of the gas, as the flexibility was included in the service offered by the supplier to its customer and its cost was rolled into the price of the gas.

In the new liberalised gas markets, marketplaces for gas (hubs) develop, where gas supply and demand is balanced by price. The various elements of flexibility are becoming distinct services to adapt supply or demand and as such have a distinct value or price of their own, determined by the market. These costs of flexibility may not be automatically rolled into the price of the gas delivered.

To provide flexibility, gas suppliers resort to flexibility tools designed to meet the varying and fluctuating volumes demanded hourly, daily, monthly and annually. Three main categories of flexibility tools are available:

- Providing large enough supply capacity to meet the highest possible demand. That can be done by own investment into production and transportation capacity or by contracting the respective production and transportation capacity;

- Providing of storage capacity in underground storage or LNG tanks to balance variations in demand;

- Arrange for customers to reduce or stop their offtake of gas at the request of the supplier according to a contractual scheme. In return the customer gets a more favourable gas price.

Gas suppliers will optimise their portfolio of flexibility instruments to meet the variations and peaks in demand. The objective is to guarantee gas supply under extreme weather conditions notwithstanding the failure of a major supply source or another system component. As neither production nor transportation capacity can be increased at short notice, surge or excess capacity needed must be anticipated at the time of the investment decision. The cost of providing such capacity is largely independent of its usage, i.e., whether used or not the same costs will be incurred.

Rising gas demand in most OECD countries has brought greater reliance on more remote supplies and reduced flexibility in supply capacity. Whereas imports from external suppliers represented 8% of OECD gas demand in 1980, they reached 20% in 2000 and are expected to reach 26% by 2010 and 32% by 2030. In view of the small share of transportation costs in the value of the gas, short-haul gas easily provides supply flexibility able to match variations in demand. In the case of long-haul gas, however, as transportation costs of long-distance pipelines constitute a large share of the value of gas, infrastructure utilisation close to the maximum is extremely desirable. This limits supply flexibility. Thus, with an increased share of long-haul gas, more storage capacity and more truly interruptible customers are needed to allow meeting variations in gas demand.

DIFFERENT FLEXIBILITY REQUIREMENTS BY SECTOR

Flexibility requirements vary greatly from sector to sector. Residential and small commercial[2] consumers experience large seasonal variations and generate a large requirement for flexibility and "swing"[3], as their consumption is mainly for heating purposes. These customers cannot easily switch to other fuels. When they install gas-fired equipment, they become "captives" of their choice for the time it would take to install alternative equipment. Generally, they are economic captives for the 15-to-20-year lifetime of the equipment and are extremely vulnerable if the gas industry is unable to provide the flexibility they require. These customers' demand can be characterised as highly price-inelastic[4]. Their demand shifts with temperature in the short-term and can increase rapidly when outside temperature drops. Their demand does, however, display some degree of price elasticity in the longer term, depending on the economic conditions of installing alternative heating equipment. Residential customers in the United Kingdom, for example, responded significantly over time to the gas price rises of the 1970s and early 1980s.

There are also non-interruptible consumers in industry, typically when gas is for process purposes. For example, when the heat from burning gas is directly

2   A distinction must be made between big and small commercial customers. Small commercial customers do not usually have the possibility to switch to alternative fuels, but big commercial customers, like airports or hospitals, generally have some back-up fuel and fuel-switching capability and a different gas demand profile.
3   The swing is the maximum gas monthly delivery divided by the average monthly delivery in a given year.
4   It should be noted, however, that residential customers can and do respond to price changes (by overheating rooms in the case of low prices, for instance). But the fact that the price of gas to these customers tends to be averaged and lagged limits their ability to respond to short-term price fluctuations.

applied to a product, the process is sensitive to any interruption, as in making sheet glass. The advantage of gas over alternative fuels, which could be stored at the customers' site, is its more uniform quality and its easier and more tailor-made handling. The non-interruptible industrial customers require a steady and reliable gas supply, but their demand is not influenced by exogenous factors as is demand for heating. These customers' demand is price inelastic in the short-term, but is stable and varies only little with outside temperature. This is also the case in industries that use gas as a raw material, notably in the production of fertilisers. If the price of interruptible gas is sufficiently attractive, petrochemical companies may store their end product in order to provide insurance against a gas supply interruption.

Demand from some industrial and larger commercial customers is different. There are a number of energy sources that these customers can substitute for natural gas. Oil and natural gas liquids can both be used as feedstock in the petrochemical industry. Both oil and coal can be used to produce steam in the industrial and electricity-generation sectors. As a result, some large gas users have installed equipment which gives them the capability to switch quickly between these fuels, depending on price and availability. They will optimise their use of two or more fuels according to operating costs, principally fuel costs. In case of power generation based on gas, shutting down the power plant and taking electricity from the grid may be an alternative.

The ability to substitute one fuel for another varies considerably across IEA regions and industries. In many cases, it is possible to switch between fuels only for short periods or with considerable investment and a time lag. Among IEA countries, the short-term fuel-switching capability from gas into oil by power generators and industrial customers is limited at 3.5 mb/d[5], corresponding to approximately 490 mcm/d. It is concentrated in only five countries: the United States, Japan, Korea, Germany and Italy. In most other IEA countries, switching capability is very small and available only for very short periods. Stocks of back-up fuel are not usually kept. Moreover, the alternative fuels for generators or industrial customers are likely to be more polluting than gas. With increasingly stringent environmental rules, switching from gas to oil may no longer be a real option. For these and other reasons, the quality of interruptible contracts varies considerably.

---

5   This figure indicates the maximum possible fuel-switching capability (IEA 2002f).

With the opening of the electric markets, the pattern for using gas in power generation has changed. Surplus power capacity can now be traded via the grid between different players and optimisation of power plant dispatch is no longer restricted to predetermined sets of alternatives under the control of individual players – for example, a utility having a portfolio of its own power plants. Electricity taken from or fed into to the grid has become a new alternative price benchmark for each player. In addition, the existence in some IEA countries of non gas-fired power plants, which have been mothballed, indirectly provides additional flexibility to the gas market. In some cases, these can be reactivated, thus freeing gas from power generation for use in other purposes.

As new trading patterns and possibilities emerge in electric markets, the time periods for cost-optimising fuel switching become much shorter. Depending on price differentials and the resulting arbitrage possibilities, additional gas demand may be induced. But the new patterns will also allow a single customer to switch much more easily away from gas, as he has the whole electricity system as a back-up. In general, liberalised systems are becoming increasingly responsive to short-term price incentives.

With the increasing development of combined-cycle gas turbines for power generation, fuel-switching capability will become more limited. Dual-firing CCGT units requires distillate rather than residual fuel-oil for back-up. Switching from gas to distillate requires a much higher natural gas price to trigger it. According to James Jensen's estimates[6], switching to distillate only became effective when natural gas prices approached $6/MBtu during the US gas shortages in the winter of 2000, whereas switching to residual fuel oil became economic when gas prices reached $3.60/MBtu.

CHANGES THAT COME WITH MARKET LIBERALISATION

With the opening of the gas sector, gas hubs have developed in marketplaces where gas is traded like other commodities. Spot and future markets develop at many hubs, where gas supply and demand are balanced by price. Some of the most liquid marketplaces are used as a reference for futures trading in gas.

Gas suppliers have begun to offer a more varied range of gas related services instead of the uniform product they offered before. Supply flexibility, for example, is offered through separate services which are valued and traded by such marketplaces.

---

6   Jensen J. 2001a.

While for some sectors, mainly residential and commercial, variation in demand must be met by gas suppliers, customers in other sectors, mainly power generation and industry, can change their demand in reaction to market prices. With the opening of electricity sectors in IEA countries there is increasing use of price arbitrage between electricity and gas markets.

New flexibility options are also appearing in LNG markets because of an increase in world-wide LNG spot trade over the last few years. On the demand side this development was stimulated by access to the highly liquid US gas market. On the supply side the costs of liquefaction plants have come down substantially, allowing gas suppliers to sell only a part of their capacity under long-term contracts while selling the remaining capacity as spot cargoes.

In liberalised markets, most gas supply and demand is balanced by trading in domestic markets or LNG spot purchases. But a still significant number of non-interruptible customers will continue to need the flexibility instruments to meet variations in the volume of demand.

2

SUPPLY AND DEMAND TRENDS IN OECD REGIONS AND THEIR IMPACT ON FLEXIBILITY

INTRODUCTION

In most OECD countries, natural gas demand is expected to increase strongly over the next three decades and account for a growing share of primary energy consumption. OECD gas demand amounted to 1,393 bcm in 2000 and is expected to reach 2,449 bcm in 2030, an increase of 1.9% a year over the period[7]. Growth in gas demand is projected to be strong in all the OECD regions. It will be strongest in OECD Pacific, with annual rates of about 2.3%, while European consumption is expected to increase by 2.1% and OECD North America by 1.7%. In all regions, the strongest increase is expected to occur in the power generation sector. The reasons for this are: the low construction and operating costs of gas-fired power plants; their high thermal efficiency; the possibility to build new gas-fired power capacity in a modular way when and where it is required (which makes gas a good fit with liberalised electricity markets); and the clean-burning properties of gas compared with other fossil fuels.

The impact on load factor of gas use in the power sector differs depending on whether the gas is used for base, middle or peak load[8]. If it is used for base-load, the overall load factor of dispatched gas will be increased. If it is used for middle or peak load, the load factor of gas will decrease. The merit order[9] of gas-fired CCGT plants in a given market will be a function of the size of the market, the fuel-mix of existing generators, the price of competing fuels and the nature of the demand profile. For instance, when LNG first entered the Japanese electricity market in the 1970s, the competitive target was base-load generation by oil, and so LNG was used for base-load generation. This is no longer true. Today the competitive targets include coal, nuclear power and hydropower. This has forced CCGT units to move increasingly into mid-load service. Now Japanese CCGT units operate at close to a 50% load factor, accentuating the seasonality of LNG demand.

---

7   This chapter draws on WEO 2002 (IEA 2002a), WEO 2001 (IEA 2001a) and WEO 2000 (IEA 2000a).
8   This paragraph and the next are drawn from a presentation by James T. Jensen on "Comparative energy costs in power generation" (Jensen J., 2001b).
9   Ranking in order of which generation plant should be used, based on ascending order of operating cost together with amount of energy that will be generated.

The fluctuations of power generation loads pose a challenge to gas supply. These daily fluctuations are usually manageable when LNG is used, but less so when the fuel is pipeline gas. Shifts in electricity loads are instantaneous. The gas transmission grid can provide some intra-day flexibility but its peak-hour delivery capacity is usually limited to some percentage of its maximum daily quantity. In the United States, for example, intraday flexibility is typically limited to 6% of the daily contractual quantity in any one hour. In some cases, it may be profitable to operate newer gas pipelines designed to carry gas-fired power-generation loads at very high pressures in order to maximise line-pack. The Yacheng pipeline that serves Hong Kong from offshore Hainan is an example of such a design.

In an environment that allows for arbitrage between the electricity and gas markets, it may be rational for gas-fired power plants to stop taking gas during gas peak-load hours, and instead to sell it on a gas spot market. The loss in power production would be replaced by electricity purchase from the electricity spot market (when possible). In such cases, flexibility in the electricity sector may enhance flexibility in the gas sector.

Flexibility is even more important in responding to predictable increases in gas use in the residential and commercial sectors. These sectors are typically responsible for the largest seasonal variations in gas demand. Gas demand in these sectors in OECD countries is expected to grow a modest 130 Mtoe (155 bcm) between 2000 and 2030, for an average annual rate of 0.9%.

The increase in demand in the residential and commercial sectors will add to the volume of price-inelastic demand. By contrast, the increase in gas demand in the power sector is rather price elastic in the short-term, due to generators' fuel-switching capacity and the back-up by the electricity grid. With the opening of the electricity market, this trend will intensify due to the arbitrage possibilities between gas and power. Over time, as gas use increases in power generation, there will be limits to the price elasticity of gas demand for power.

OECD NORTH AMERICA

Market overview

The North American gas market is the largest in the world with 733 bcm consumed in 2001, or 29% of global gas demand. The United States accounted for 608 bcm. The region is self-sufficient in gas, with a reserves-to-production ratio in US over several decades of about 7 to 8 years. Over recent years, depletion rates for gas have been accelerating. North America has a large yet-to-be explored potential. Remaining resources are 27-34 tcm, while remaining proven reserves are about 7.8 tcm (Cedigaz), so that new gas will probably be developed when the market signals are right and if there are no regulatory obstacles.

US imports of LNG, which reached 6.75 bcm in 2001, have so far been of minor importance for the volume balance of North America. This situation is evolving. LNG is now viewed as a "backstop" supply the US can fall back on if conventional production does not develop in line with expectations[10]. For example, LNG imports increased in 2000 and 2001 triggered by high US gas prices. Numerous spot LNG cargoes complemented long-term LNG supplies to the US East Coast from Algeria and from Trinidad and Tobago.

Canada and the US have similar regulatory frameworks as both are NAFTA members, and large pipeline links carry substantial flows of gas from Canada to the United States (107 bcm in 2001). The two countries' gas systems may, in fact, be considered a single network. However, interconnection between the countries varies considerably among states and provinces. The western United States is more closely integrated with Canada than it is with the Eastern US market. Mexico is also connected to the US market, but the capacity of the connecting pipelines is small.

Regulatory reforms in gas and electricity

The US and Canadian gas industry has undergone profound structural changes over the last two decades, largely due to regulatory reforms aimed at promoting competition and improving efficiency. This process began with the lifting of controls on wellhead prices, followed by mandatory open access to the interstate pipeline and storage system and the unbundling of pipeline companies' gas-trading, transportation and sales activities. To date, open access is mostly limited to industrial and large commercial end-users. Several states and provinces are now expanding open access and retail competition to small residential and commercial consumers. Pricing of transmission services remains, for the most part, regulated by the national regulators, the National Energy Board in Canada and the Federal Energy Regulatory Commission in the United States. Distribution services come under state and provincial regulators, usually on a traditional cost-of-service basis.

---

10 Jensen J. 2002.

The US and Canadian power industries have also undergone major regulatory changes, since 1992. Unlike the gas industry, electricity generation and the regional electric grids are regulated on the state or province level. A major share of the overall reform effort has been aimed at intensifying competition between power generators, mainly through the provision of non-discriminatory access to the transmission grid. New regulation focuses on levelling the playing field for supply competition by means of unbundling and by transparency obligations imposed on the utilities. On the end-user side, reforms aim at enabling all consumers to choose their supplier and at strengthening consumer protection. In some states, retail electricity customers can now choose their electricity supplier[11], but the California power crisis has slowed the pace of reform in a number of states.

Wholesale electricity markets, which only recently came into existence, are now operating in many parts of North America. The emergence of centralised power markets has significantly changed the way power is sold. Numerous electricity trading hubs[12] have emerged over the past few years, as was the case in the gas sector during the 1990s. There are currently ten major trading hubs for electric power in the United States, five of them are located in western US, four in the Midwest, and one in the East. The NYMEX (New York Mercantile Exchange) and the CBOT (Chicago Board of Trade) have developed and sponsored electricity futures contracts to facilitate trading at these hubs. The existence of futures contracts has increased the linkage between the gas and power sectors, facilitating arbitrage between the two commodities.

In Mexico, the government announced plans in 2000 to restructure the energy sector and reform its legal and institutional framework to boost efficiency and investment. The reforms would grant operational and financial autonomy to the public energy companies and would lower their high tax burdens to enable them to reinvest. The role of the private sector would be enlarged. Main elements of the planned reforms that are relevant to natural gas and electricity are:

- Private oil and gas companies will be invited to tender bids to develop non-associated gas fields on behalf of Pemex under long-term service contracts.

- Private and foreign companies will be allowed to participate in building and operating LNG import terminals.

---

11 EIA (2000a).
12 A hub is a location on the power grid representing a delivery point where power is sold and ownership changes hand.

- Independent power producers will be allowed to sell directly to end-users any power not bought by the Federal Electricity Commission.

- Energy subsidies will be phased out, and all fuels will be priced on the basis of full supply costs.

Gas demand trends

OECD North American gas consumption has been rising steadily since the mid-1980s, from 579 bcm in 1985 to 733 bcm in 2001. Most of the increase in US demand has been met by Canadian production (70% of the increase over the period 1985-2001). Canadian and Mexican demand have also increased rapidly, from 61 bcm in 1985 to 85 bcm in 2001 in Canada and from 28 bcm in 1985 to 39 bcm in 2001 in Mexico.

US gas consumption decreased by 6% in 2001, as high gas prices, fuel switching and lower economic activity, especially after the 11 September terrorist attacks, reduced industrial and power-sector demand, while deliveries to commercial and residential consumers were low in response to above-normal temperatures. Gas consumption started to rise again in 2002 in line with US economic activity.

North American gas consumption is projected to rise at an average annual rate of 1.7% between 2000 and 2030, from 788 bcm to 1,305 bcm. Its share in total energy supply will increase from 24% in 2000 to 29% in 2030. Power generation accounts for three-quarters of the increase. During the past five years, demand growth for gas in electric power has been particularly strong in the United States (11% per year between 1996 and 2000). The gas-fired share of US electricity generation, including co-generation, rose from 13.2% in 1996 to 16% in 2000. About 22 GW of new gas-fired generating capacity was added in 2000 and 40 GW in 2001. According to the Utility Data Institute, this trend will continue over the next five years. Of 253 GW of generating capacity planned or under construction in the United States, 238 GW will be fuelled by gas. Electricity generation is therefore expected to surpass the industrial sector as the largest consumer of natural gas in the United States.

Gas in power generation in the US is principally used for middle load. The new capacity recently built, under construction or in planning consists of 170 GW of CCGT (typically used for base load), 66 GW of gas turbines (typically used for peak load[13]) and 2 GW of cogeneration. New gas turbines that can

---

13 Generating units that can be brought on line quickly and used to meet requirements during the greatest or peak-load periods on the system (for instance, during heat waves on the East Coast).

be switched on and off at short notice are well suited for arbitrage between the power and gas sectors and are already causing substantial swings in gas requirements.

Gas supply trends

Proven gas reserves in North America amounted to 7.8 tcm at the beginning of 2002, 4% of global gas reserves. Just above two-thirds of these reserves are in the United States. Production is concentrated in the southern and central US states and in western Canada.

World Energy Outlook 2002 foresees a 27% increase in North American gas production over the period 2000-2030. Marketed production in 2030 is expected to reach 960 bcm, up from 773 bcm in 2001. North American gas production prospects depend on producers discovering and developing conventional and unconventional reserves and connecting them to markets. Higher output will require increased drilling in established producing basins in the lower 48 US states and in Canada, as well as new greenfield projects. Aggregate production in the United States and Canada is projected to climb slowly from 736 bcm in 2001 to 823 bcm in 2010 before beginning to decline around 2020, to 812 bcm in 2030. In Mexico, the long-term outlook is more promising. Gas output is projected to grow from 37 bcm in 2001 to 148 bcm in 2030.

According to the National Energy Board[14], exports from Canada to the US could reach a maximum of about 140 bcm in 2018, before declining moderately to 130 bcm by 2025. Imports of LNG will play a growing role in US gas supply in the long term. Capacity expansions at the four existing LNG facilities on the Gulf and East Coasts and potential investment in new terminals in the United States, or neighbouring countries, will depend on prices and project-development costs. The rising cost of gas from domestic sources, together with continuing reductions in the costs of LNG supply, is expected to boost US imports of LNG, directly or via Mexico. There may also be potential for large-scale imports of Mexican gas by pipeline into the United States. But this will probably not happen before 2020, as Mexico will need to meet its domestic needs first.

Impact on flexibility

Increased demand in the residential and commercial sectors, as well as the use of gas in gas turbines for peak load, will increase the seasonality of gas demand.

---

14 National Energy Board (1999).

This will be somewhat counterbalanced by the increasing use of gas in CCGT for base load and by supply trends, including an increased share of LNG in gas imports. LNG imports are more flexible than long-distance pipeline imports and could increase the swing in gas imports, reacting to price signals.

Arbitrage possibilities between gas and electric power resulting from the liberalisation of both sectors will increase the elastic part of gas demand, as operators will have the possibility to sell their gas on the spot market and buy electricity from the power exchanges or vice versa. The result, as can already be seen today, will be an increased use of flexible instruments, such as high-deliverability gas storage.

OECD EUROPE

Market overview

With 490 bcm consumed in 2001, OECD Europe is the third largest regional gas market after North America and the Former Soviet Union. European gas demand grew at an average rate of 3.7% per year from 1973 to 2000. The residential sector is the single largest consuming sector, followed by the industrial, electricity and commercial sectors. The use of gas in power generation is growing rapidly, especially in the United Kingdom. The largest gas markets are the UK, Germany, Italy, the Netherlands and France. Together, they accounted for three-quarters of OECD European gas consumption in 2001.

Indigenous production, concentrated in the UK, Norway and the Netherlands, has grown in recent years, but not fast enough to keep pace with demand. Imports from external suppliers have therefore increased, and now account for 41% of total European gas needs, compared to only 17% in 1980. Russia and Algeria are the two major external suppliers. Inside the region, Norway and the Netherlands provide the bulk of internal trade.

The Continental European gas market is characterised by a small number of large gas companies, most of them with the state as a major shareholder, and with strong positions in their home market. Legislation at national and European level to liberalise gas markets by introducing third-party access and unbundling have brought a degree of gas-to-gas competition and have fostered spot markets. By contrast, the UK market is fully private and liberalised and UK has its own substantial gas reserves.

In Continental Europe, price signals do not necessarily work their way to the producing countries, especially to Russia and Algeria, where upstream investment decisions are triggered by many other factors than market prices in the European Union. This is not the case so far in the UK where supply reacts quickly to price signals from the market.

Regulatory reforms in gas and electricity

The pace of market reform varies widely across Europe. Gas-to-gas competition is most developed in the UK, where structural and regulatory reforms were first launched in the late 1980s.

In Continental Europe, market liberalisation started in August 2000 with the transposition of the EU Gas Directive into national laws. The directive established common rules for the implementation of competition in the gas sector, through third-party access (TPA) to the transmission networks. The directive allows either "negotiated" TPA with the publication of the main commercial conditions, or "regulated" TPA, based on published tariff structures. Most EU member states, except Germany, have opted for, or are moving to, regulated TPA. National approaches to unbundling of transportation and supply activities are quite mixed. Some countries are moving towards organisational unbundling, rather than unbundling of accounts. In 2001/2002, a number of companies in Belgium, Spain, Italy and the Netherlands decided to separate their transmission and supply activities.

Competition was introduced in all EU member states' electricity markets in 1999, when the Electricity Directive was transposed into national laws.

In the UK, gas and electricity markets have been fully opened since 1998. All households are eligible to choose their supplier. In October 1999, New Gas Trading Arrangements (NGTA) were published, introducing entry capacity auctions. Under New Electricity Trading Arrangements (NETA) implemented in March 2001, the old mandatory electricity pool has been replaced by a system in which electricity trade is mostly bilateral contracts.

In March 2002 the Council of the European Union agreed to vote on new EU gas and electricity directives before the end of 2002. The new draft directives foresee an extension of the right to choose their gas and electricity suppliers to all non-household consumers by the end of 2004. This will amount to at least 60% of the total market. The new directives were tabled by the EU Commission in 2001. They aimed to open electricity and gas markets fully, to reduce existing barriers to competition and to reinforce the regulatory function.

Gas and electricity market reforms have led to the emergence of new market places, in addition to APX in Amsterdam; Nordpool in Oslo; OMEL in Madrid; and Elexon, IPE, NGC, PowerEx and the UK Power Exchange in the United Kingdom. PowerNext opened in France in November 2001 and an Italian power exchange is expected to start operation in 2003[15]. The two German power exchanges, located in Frankfurt and Leipzig, merged in March 2002. On the gas side, a new hub is developing at Bunde in Germany, in addition to the two existing hubs, the National Balancing Point in UK and Zeebrugge in Belgium.

Gas demand trends

Gas consumption in OECD Europe is expected to continue growing faster than that for any other fuel, at an average rate of 2.1% per year between 2000 and 2030, resulting in 87% growth from 482 bcm in 2000 to 901 bcm in 2030. Gas would then become Europe's second fuel after oil, with 33% of TPES in 2030 (22% in 2000).

Gas use is increasing in power generation and in all end-user sectors. Seventy-two per cent of the expected increase in European gas demand by 2030 is explained by increased use of gas in the power generation sector, where gas will take up the predominant share in market growth. The large European increase of gas in the power sector has been led by the dash for gas in the UK, where gas is used for base-load power generation. The relative prices of power and gas in the UK under the new NETA regime are such that gas-fired CCGT plants cannot recover their full costs. Gas use may therefore be reduced. It already happened in 2001 in UK when generators sought to cut their fuel costs. The extent of the reduction in gas use will depend on whether environmental considerations restrict coal use and on the possibility of importing electricity from neighbouring countries.

A limited increase of 55 Mtoe, or 66 bcm (+ 1% per year) is expected over the period 2000-2030 in the residential and commercial sectors, while consumption by industry is expected to increase on average by 0.8% per year during the period.

---

15 IEA (2002e-forthcoming).

Gas supply trends

OECD Europe's 7.35 tcm of proven gas reserves at the beginning of 2002 represents 4.5% of the world total (Cedigaz). About 80% of the region's reserves are in Norway, the Netherlands and the United Kingdom. For several years, the reserves-to-production ratio has been stable at about 20 years, despite rising production in the North Sea. Major reserves have been added mainly by new finds in the North Sea and Norwegian Sea and by higher recovery rates in existing fields, due to techniques to improve recovery, mainly horizontal drilling. European gas production amounted to 308 bcm in 2001. Norway exports substantial volumes of gas to the Continent and small volumes to the United Kingdom. The Netherlands export half of their gas production to other European countries. The United Kingdom has become an exporter of gas to the Continent since the opening of the Interconnector in October 1998, but due to price differentials across the Channel, the UK sometimes imports gas, as the Interconnector can reverse its flows.

Most European countries depend heavily on gas imports from distant sources and from a limited number of producing countries, in several of which exports are exclusively handled by state monopolies. Russia is the largest external supplier to Europe, providing 117 bcm or just under two-thirds of total European imports from external suppliers, and a quarter of total supply – entirely by pipeline. Algeria, with 55 bcm, is the next biggest exporter of gas to Europe, both via pipeline and as LNG. Imports of LNG play a small, but increasing role in OECD Europe. LNG imports from Nigeria began in 1999 and from Trinidad and Tobago in 2000. Europe has been importing small volumes of LNG from Libya since the early 1970s and spot cargoes from the Middle East in recent years.

Gas production in OECD Europe is expected to remain stable at approximately 300 bcm a year until 2020 and then decrease slightly to 276 bcm in 2030. Production could turn out to be higher depending on technological and price developments. Nevertheless, given the limited gas resources in most European countries and the prospect of rising demand, imports from Norway and from outside the region are expected to continue to increase for the next two decades.

Incremental European gas import requirements are likely to be met by increased piped supplies from the two main existing suppliers, Russia and Algeria, and a mixture of piped gas and LNG from other existing or emerging suppliers. The latter will probably include Libya (via pipeline), with LNG coming from

Nigeria, Trinidad and Tobago, Egypt and Qatar. Venezuela may also emerge in the longer term as a bulk supplier of LNG, while shipments of LNG from other Middle East producers (whose reserve potential is large and still little used) may also increase. LNG, both under long-term contracts and via spot purchases, could play a much more important role in supplying the European gas market.

Currently, most gas in Continental Europe is imported through long-term Take-or-Pay contracts. These contracts divide the risks associated with highly capital-intensive gas projects among producers and importers. They typically put the price risk on the seller and the volume risk on the buyer (ToP obligations). With market reform some significant changes in the structure and pricing of these long-term ToP contracts will be observed.

Experience in liberalised markets shows that long-term contracts do not entirely disappear with market liberalisation. In the United Kingdom, about 85% of the gas delivered at the beach is covered by contracts, as is about half of the wholesale gas in the United States. But the eight-to-ten-year duration of US contracts is shorter than in continental Europe and contracts have different pricing rules. Moreover, gas is traded repeatedly before it reaches the burner tip, and there are numerous renegotiations between producers, suppliers and consumers on price, volume and delivery point. As gas-to-gas competition increases and oil products become less competitive with gas in the power generation markets, prices in most contracts are partly or totally indexed with changes in spot or futures gas prices.

Some changes can be observed in new continental gas supply contracts as a reaction to market reform and as part of a portfolio approach by buyers: shorter terms for new contracts (8-to-15 years instead of 20-to-25 years), smaller volumes and greater flexibility in reviewing contractual terms, new price indices including electricity pool prices and spot gas prices. On the Continent, however, some very large long-term ToP contracts will remain in force for decades to come. These include contracts between Gasexport and Ruhrgas (prolonged in 1998 until 2030), the Troll contracts, Gasunie and Sonatrach contracts. These contracts are less rigid than older Asian LNG contracts, as they include regular renegotiation clauses.

Box 1: Flexibility in Long-Term Contracts

Long-term Take-or-Pay (ToP) contracts are generally considered to be "rigid" contracts, mainly because they link sellers and buyers for 20 to 25 years, over which both of them have strictly defined obligations. They include a ToP clause, under which the buyer must pay for a certain amount of gas whether taken or not, as well as an obligation on the seller to make available defined volumes of gas. Such contracts constitute a firm basis for both sellers and buyers to finance a highly capital-intensive infrastructure.

In some contracts price conditions are fixed for the lifetime of the contract, others provide instruments to renegotiate prices to adapt them to changed circumstances. The first type has been usual practice in UK and US; the second, in Continental Europe.

Experience with long-term ToP contracts in the UK and the US indicates that the contract obligations can be very costly for the companies involved when circumstances such as the regulatory framework change substantially and there is no re-opener clause. In the United Kingdom, it cost British Gas £2.5 billion to renegotiate its contracts with North Sea producers when gas prices dropped to 9 p/therm, while the cost to British Gas for buying it averaged 19 p/therm.

Long-term ToP contracts in Continental Europe allow for regular price renegotiations every three years by which price and in some cases other conditions can be revised if market circumstances change significantly. If the parties cannot agree on changes, either party may go to arbitration but this has rarely happened.

Long-term contracts in Continental Europe usually provide that the pricing provision should reflect the competitive situation in the buyer's market. This approach results in different gas price formulas – and prices – to the market of different buyers. These price differentials create an incentive for buyers to profit from price differentials by reselling part of their offtake to more lucrative markets. While sellers tend to try to restrict the use of such price differentials by their buyers, any restriction of the use of gas to the country of the buyer (destination clause) would certainly not be in line with European law.

Long-term ToP contracts include an element of volume flexibility, which depends on the economics of the production and transportation side. Short-haul gas from pure gas fields is offered with a large swing and large flexibility while long-haul gas or offshore gas is offered with more limited swing and flexibility. Gas from associated fields tends to carry most rigid offtake obligations to avoid flaring. The value of the flexibility offered by the contract is reflected in the price of the gas.

The flexibility offered by long-term contracts consists mainly in the difference between the capacity and the off-take/minimum-pay obligation for various time spans; a year for long-term flexibility and a day or an hour for short-term flexibility:

- Annual flexibility is typically of the order of 20%. The minimum pay-out is mainly a provision to protect the cash flow of the seller. So, contracts usually have a carry-forward provision, which allows gas taken in excess of the minimum pay obligation to be credited against future minimum pay-out obligations. There is also a make-up provision, under which gas not taken may be treated as a prepayment which can be counted against future off takes above the minimum pay obligation. In addition there may be clauses that adjust the annual volume obligations as a function of exogenous influences like average temperature.

- Daily flexibility is measured by the difference between daily capacity and the daily off-take obligation. A daily off-take obligation is not necessary to protect the seller's cash flow. But it may be necessary because of technical minimum-flow restrictions or to guarantee the outlet of associated gas. Gas from pure gas fields usually has a very low daily minimum off-take obligation or none while gas from associated gas fields may have a minimum daily off-take obligation not very much lower than the one-day average annual obligation.

- The "swing" given by a contract results from the relation between the daily and annual obligations. Where the daily availability is considerably higher than the average annual availability the contract allows for a large swing to follow the market. The Dutch contracts provide such a large swing; so did the early UK delivery contracts from the fields in the southern UK Continental Shelf. Both were for short-haul gas. For long-haul gas, daily availability is usually not larger than average annual availability; as a higher daily availability would require investment in capacity without a guaranteed cash flow.

- A new feature was included in some of the long-term UK export contracts for delivery to the Continent via the Interconnector. It is called a "clawback gas" provision. It gives the supplier the right to interrupt the normal contractual flow to take advantage of high UK spot prices. Contrary to the traditional volume-related elements of flexibility this adds a price element to the flexibility provisions.

Source: ESMAP (1993), European Gas Matters, 30 April 2002.

Impact on flexibility

On one hand, increased use of gas in the residential and commercial sectors will increase gas demand seasonality in Europe. On the other hand, the largest increase in gas demand is expected from power generation, which might reduce gas seasonality if the gas is used for base-load generation. This has been the case in the United Kingdom so far.

European gas companies have recently begun to increase storage capacity in order to meet the challenges that a temperature-sensitive market may present during the next few decades. At the beginning of 2000, the working gas volumes of European storage facilities amounted to 59 bcm (13% of 2000 gas consumption) compared with 31 bcm at the beginning of 1990 (12% of 1990 gas consumption).

For the time being, oversupply and some spare capacity in transportation systems provide a great deal of gas flexibility on the supply side, although some transmission pipelines are already operating near capacity. Increasing demand will gradually reduce spare capacity and decrease the current flexibility, spurring increase in storage capacity and interconnectivity of the gas grids. Supply and transportation flexibility will also decrease as imports increasingly come from more remote areas and exporters seek to make best use of their assets. This will be partly compensated by an increase in LNG imports, which are more flexible and could increase the swing in overall gas imports.

On the other hand, market liberalisation will offer new ways to deal with flexibility requirements: swap agreements, spot markets, LNG spot cargoes, more flexibility in contracts. Liberalisation will also bring market-based valuations of the various forms of flexibility. Improved interconnection of the gas transmission network in Europe will be an important factor in enhancing flexibility and security. Because of the diversity of OECD European markets, improved interconnection can allow additional short-term gas to be provided from neighbouring markets, if there is a large difference in prevailing demand, such as might happen with cold weather in one market and milder weather in a neighbouring interconnected area.

Since the United Kingdom was geographically isolated from the Continental European gas grid and the country was self-sufficient in gas, the UK gas market developed in its own way, which was more like the US gas market than that of Continental Europe. As gas production on the UK Continental Shelf is not keeping step with gas consumption, the UK market could be in a similar situation to that of Continental Europe in the next five-to-ten years, with a growing share of imported gas and the need to find additional flexibility.

OECD PACIFIC

Market overview

OECD Pacific consumed 131 bcm of gas in 2001. The region's main gas markets contrast sharply. Japan and Korea, with limited reserves of their own, are large LNG importers, whilst Australia is a large LNG exporter. Gas use has grown very rapidly in the region (+ 11% per year during the period 1971-2000) due to massive LNG imports by Japan and Korea.

Natural gas resources and production in OECD Pacific are concentrated in Australia, with a small amount in New Zealand, which is self-sufficient. Australia is a major LNG supplier to Japan. But the main part of the region's gas needs is met by LNG imports from the rest of East Asia/Pacific and the Middle East. Imports, which account for 96% of natural gas use in Japan, are entirely in the form of LNG, coming from Indonesia, Malaysia, Australia, Brunei, the United Arab Emirates, Qatar, Alaska and Oman. Imports reached 78 bcm in 2001. South Korea imported 21 bcm of LNG in 2001, mainly from Indonesia, Malaysia, Qatar and Oman.

Demand is dominated by power generation which used 54% of the region's gas in 2000. Demand and trade – both within and outside the region – will continue to grow at different rates across countries. In New Zealand, gas consumption is set to decline in line with the depletion of national gas reserves; little growth is expected in Japan, where the major uncertainty lies in the amount of gas that will be consumed in future by the power electricity sector. Strong growth in consumption is expected in both Australia and Korea.

Regulatory reforms in gas and electricity

Downstream markets in Japan and South Korea are in a state of change, as governments introduce competition into the electricity and gas markets.

In Japan, retail power sales to large users who account for about one-third of Japan's power demand were opened to competition in March 2000. As a result, competition is emerging among existing players and new entrants and Japanese power companies have become much more cost-conscious than before. Further reforms of the electricity sector will include the unbundling and regulation of transmission, the creation of a spot market and the opening up of the retail market[16].

---

16 IEA (2002e-forthcoming).

The gas industry is undergoing changes as a result of power-sector deregulation and of emerging competition in domestic gas distribution. Amendments to the Gas Utility Law passed in 1999 and 2000 liberalised gas sales to industrial customers whose annual consumption exceeds one million cubic meters. The government has speeded up the access of new entrants to the existing large-user market, by dropping a requirement for prior approval, in favour of a mere notification of intention to compete. Existing gas and electricity utilities, and new entrants, mainly oil companies, can now sell in one another's territory. Some Japanese gas and electricity utilities are beginning to look outside their traditional markets, and some oil refiners are looking at new opportunities in the LNG retail market.

Tokyo Electric Power Company (Tepco), Japan's major importer of LNG, is aiming to become a wholesale gas supplier. It has signed a contract with Keiyo Gas, Japan's fifth-largest local gas distributor. Tepco is also selling gas to Otaki Gas, another Chubu-based gas distributor. Tepco plans to start selling surplus LNG in 2006 under two-to-three-year contracts with small domestic gas companies.

Tokyo Gas, the largest gas utility company, has recently established a joint venture with Shell, Nijo Gas Co, to sell gas to independent power producers and other large consumers in Tokyo's metropolitan district.

The Japanese Ministry of Economy, Trade and Industry (METI) is taking steps to ensure competitive third-party access for small- and medium-sized companies to city gas grids. Third-party access to LNG facilities, storage tanks and pipelines is due in 2003. The government is also investigating how to develop a natural gas grid to foster competition.

In Korea, the electricity and gas sectors are in the early stages of a long restructuring, privatisation and liberalisation process. In April 2001, non-nuclear power-generation assets were divided into five generating subsidiaries still wholly-owned by the state, but with privatisation planned to begin in 2002. A cost-based electricity pool and power exchange was put into operation in April 2001 and is expected to be transformed into a competitive mandatory pool in 2003. By the end of 2002, distribution subsidiaries are to be created. In the gas sector, gas imports and sales are going to be unbundled from terminal and transmission operations and the government intends to open access to LNG receiving terminals and the transmission network in 2003.

In Australia, electricity reforms have occurred at both state and national levels. At national level, the aim was to develop a national electricity market (NEM) for the wholesale supply and purchase of electricity in five Australian states and territories. The market provides open access to transmission and distribution networks by generators, retailers and customers, co-ordinated planning of the interconnected power systems in the NEM jurisdictions and the maintenance of system security. At state level, reform was led by the two most populous states: Victoria, which privatised and restructured its electricity sector in 1994, and New South Wales, which established a daily electricity pool in 1996.

The Australian natural-gas market has also changed considerably in the last few years. Competition has been introduced through non-discriminatory open access to pipelines and intensified interstate trade through the removal of regulatory barriers and the interconnection of state infrastructure.

In New Zealand, a package of reforms to promote competition in electricity generation, distribution and retailing was adopted in 1998. Non-discriminatory access to the gas transmission grid was introduced in 1999.

Gas demand trends

Natural gas consumption in OECD Pacific reached 131 bcm in 2001. It is expected to grow at an average annual rate of 2.3% from 2000 to 2030. This is far less than the 11% growth observed during the period 1971-2000. Gas consumption is expected to increase from 122 bcm in 2000 to 243 bcm in 2030.

In Japan and Korea, the gas market was shaken by the Asian financial crisis of 1997-1998, when LNG buyers found themselves vastly over-supplied. Growth has picked up again and is expected to continue. However, the pace of future growth remains uncertain. In Japan, future demand will depend on the strength of the economy and the power sector, which consumed approximately 70% of LNG imports in 2000. The gas requirements of electric power companies are not expected to grow much in Japan, due to competition from coal and nuclear power, and to low economic growth generally. On the other hand, demand for residential and commercial use, which is relatively inelastic to the economic situation and prices, is expected to increase rapidly. In the 2001 review of Japanese Energy Policy, the Japanese government foresees an increase in total gas consumption of 7 Mtoe or approximately 8 bcm (10% in all) over the period 2000-2010 under its "policy case" scenario (which includes planned policy measures to achieve Japan's Kyoto targets).

Until recently, LNG consumption was relatively stable throughout the year as power plants used it for base-load generation. In the past few years, however, LNG consumption has become increasingly seasonalised, with two peaks: one in summer induced by electricity needs for cooling and one in winter for heating. Residential and commercial sector consumption has been growing fast in Japan and Korea and this trend is expected to continue. In addition, because of competition with coal and nuclear power, gas-fired power plants are increasingly being used for middle and peak loads. This is expected to continue in the future. In Japan, the electric utilities will use soon LNG as a swing or buffer fuel, much the way oil is used today[17].

In Korea, where seasonal fluctuations in gas demand are high, the Korean gas company KOGAS gives financial incentives for gas-operated cooling systems hoping to smooth demand by promoting gas sales in the summer season.

In Australia, most demand growth is expected to come from the power generation and industrial sectors.

Gas supply trends

Australia has large and increasing proven gas reserves, 3.6 tcm as of 1 January 2002, according to Cedigaz, and a large resource potential. The country will provide the bulk of the future increase in OECD Pacific production, which is expected to grow from 42 bcm in 2001 to 63 bcm in 2010 and 122 bcm in 2030. The dependence of OECD Pacific countries on gas imports from outside the OECD is expected to fall, as Australia will partly provide the growing volumes of imported LNG in Japan and Korea.

Nevertheless, both Japan and Korea will continue importing large amounts of gas from outside the region. The principal options for Japan to meet any future increase in natural gas demand include increased imports of LNG from Asia, the Middle East, Alaska or Sakhalin. There is also the potential for Russian gas piped from Sakhalin or Irkutsk via China and the Korean Peninsula. The most likely sources of additional LNG supplies are from existing major suppliers and possibly Russia. Japan is already committed to lifting more LNG from Malaysia, starting in 2003, and from Australia's Northwest Shelf expansion, starting in 2004. In Korea, the options for imports from new greenfield LNG projects include LNG from Tangguh in Indonesia, the new Australian projects, Gorgon and North Australia, Russia's Sakhalin-2 and new Middle East projects.

Impact on flexibility

Gas demand in OECD Pacific is set to become more seasonal because of increased use of gas in the residential and commercial sectors and for upper and

---

17 FACTS (2002).

middle load in the power generation sector. Increasing seasonality in gas demand will require increased flexibility of supply. This could be achieved through a combination of swap agreements between companies in the region, spot LNG cargoes purchased during peak requirements periods and increased flexibility in new long-term contracts.

Increased LNG purchases in Japan and Korea will probably be conditional on more competitive pricing formulae and more flexible contracts, with shorter terms. A growing proportion of any new demand is likely to be met by spot purchases.

3

FLEXIBILITY TOOLS IN TRADITIONAL MARKETS

INTRODUCTION

Before market liberalisation, the entire gas demand curve was largely inelastic. It was also subject to exogenous influences such as weather, focussing the overriding aim on ensuring that demand was always met. It was, in effect, a matter of volume management. The main requirement was that the available supply should at all times be sufficient to cover contractual demand at each location.

Large suppliers such as state monopolies or companies with exclusive supply concessions were meeting this requirement. Replacement-value pricing with reference to crude oil meant that gas competitiveness was linked to other fuels. As such, there was little incentive for customers to switch to alternate fuels in response to prices. Flexibility had a cost, but it was included as part of a bundled service and the costs were passed on to customers in gas prices.

In traditional markets[18], gas companies have developed several mechanisms for balancing gas supply and demand for discrete periods of time. Some of these mechanisms are supply-side techniques that adjust the volume of gas made available by the system. Others work on the demand side, by influencing the call on gas. Buffer tools, such as underground gas storage and liquefied natural gas storage tanks also play an essential role in balancing supply and demand.

An important aspect of flexibility is the timing required: hourly, daily, monthly and annually. Different tools apply according to the required timing:

- flexibility in supply (production and imports) is required to cover monthly and annual variations in demand;
- storage can usually cover daily and seasonal variations in gas demand;
- line-pack is generally used to cover flexibility requirements during the day;
- interruptible contracts are ultimately available in the case of a sudden unforeseen increase in gas demand (wave of cold weather) or in the case of interruptions in gas supply.

---

18 «Traditional markets», as used here, refers to markets where there is little or no gas-to-gas competition as opposed to fully competitive markets.

Market players' access to flexibility tools can be directly through ownership, as in the case of a vertical operator, or it may be through contractual relationships with suppliers. Access may also take the form of anonymous market instruments such as commodity exchanges in more mature markets.

FLEXIBILITY OF SUPPLY

Flexibility in gas production and imports

Certain fields play a large role in meeting seasonal demand variations. These fields are typically non-associated gas and have geological characteristics such as high pressure and high permeability as well as favourable economic parameters, such as low investment requirements for production and proximity to the market. They are, as a result, able to vary their production up and down as required. This is particularly the case of the Groningen field in the Netherlands, which is considered the swing provider for Europe (Box 2).

Flexibility at a given delivery point can be measured by the "swing", which is defined as the maximum monthly delivery divided by the average monthly delivery in a given year.

■ **IEA Europe**

For IEA Europe, the average swing at the production and import points was 135% in 2000. The two major producing countries, the Netherlands and the UK, present a high level of swing. In other producing countries in Europe, the swing is more limited. In the Netherlands, the monthly production in January 2000 was 3.5 times higher than in August 2000 (Figure 1).

Flexibility in supply contracts is an important element to meet variations in demand, but it depends heavily on economic considerations. The main fields that supply gas to Europe, such as Troll in Norway, Urengoy in Russia and Hassi R'Mel in Algeria, have high flexibility to vary their production rate. However, the high and fixed transportation costs due to long distance or offshore transportation result in a rigidly defined upper limit on the delivery capacity from those fields, which allows little flexibility. Groningen is an exception, being close to the main markets.

Generally, import contracts to the EU provide for a flexibility allowance of 90-110% of contracted volumes. The swing shown in exports from Russia, Norway and Algeria is in the range of 110-120%. Exports from the Netherlands showed

Figure 1: European Monthly Gas Production

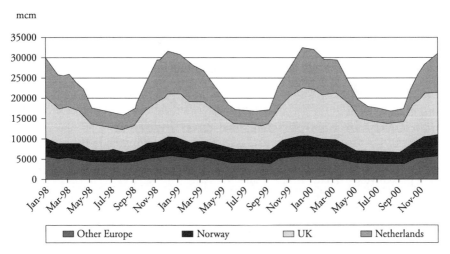

Source: IEA Monthly Database

Figure 2: European Monthly Gas Imports from the Netherlands, Norway, Algeria and Russia (1995-2000)

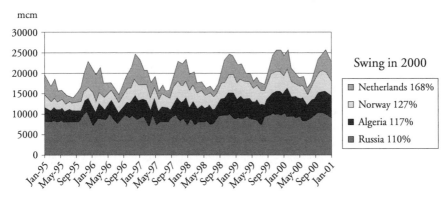

Source: IEA Monthly Database.

a swing of 160% or more, reflecting the country's role as a swing supplier. An interesting development is the increased flexibility of Norwegian gas exports[19].

___

19 Statistics on suppliers other than the Netherlands require cautious interpretation. These swings may, in fact, be quite different from the maximum swings provided for in the contract. The statistics do not show the reasons for the respective swing and may differ from the swing that might be contractually possible. They do not reveal to what extent incentives in the form of summer rebates may influence the importer's physical import pattern.

Box 2: Groningen, a Swing Supplier for Europe

The Dutch Groningen field was discovered in 1959 and began production in 1964. Of its initial 2,800 bcm of recoverable reserves, 1,600 bcm have been produced to date. Twenty-nine production clusters and 300 wells are used to produce the gas. Nederlandse Aardolie Maatschappij B.V., a 50-50 joint venture of Shell and Exxon Mobil, is the operator. Production can vary greatly and follow demand variations, according to the temperature or the business cycle, as shown in the figure below. Groningen has thus become the major swing supplier of Europe. Gas from Groningen marked the beginnings of the natural gas industry in Europe. Given its geological characteristics and its proximity to the main market areas, the Groningen field could economically provide gas with a swing to meet market requirements.

Figure 3: Groningen Production Profile in 2000

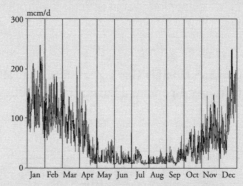

Source: NAM[20].

Groningen can be considered as an exception, however, not only because of its intrinsic "predisposition" to act as a swing producer, but also because of the policy adopted by the Dutch government in the 1970s to use Groningen to tie in smaller, less flexible fields. This would extend the life time of Groningen while providing the necessary flexibility to market the gas from those smaller fields.

Gas from Groningen is "L-gas" - a gas with a lower calorific value than "H-gas" from the North Sea and Russia. L-gas has relatively high nitrogen content. It is delivered to northwest Europe through a gas grid that is separate from the grid for H-gas and has separate storage facilities. As Groningen gas is delivered in line with market requirements for flexibility, only very little storage capacity for L-gas has been developed.

---

20   Van Nieuwland A.J.F.M (2001).

- **A decrease in UK production swing**

Virtually all domestic gas in the United Kingdom is produced offshore in the North Sea and the Irish Sea. Traditionally, British gas was produced with a high swing factor, averaging from 150% to 160% up to 1995. Output from the Morecambe field landed at Barrow had the highest swing among the major fields. The swing of gas production in the United Kingdom has declined since 1995, to just 124% in 2000.

The decrease in the swing factor has several causes. The increasing use of gas for base-load power generation has decreased the percentage of demand requiring swing. There has also been an increase in the output of gas in association with crude oil or condensates. Swing is generally much lower for this sort of gas because producers have a strong incentive to keep oil production constantly high. Finally, with the removal of the British Gas' monopoly, producers have also been lowering swing to improve profits.

Figure 4: UK Swing Gas Production

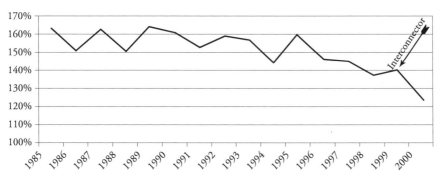

Source: IEA Monthly Database.

The recent sharp decrease in swing production can be partly explained by the balancing role played by the UK-Belgium Interconnector, which entered into operation in October 1998 (see Annex on United Kingdom). The decrease in swing production also demonstrates the rather high cost of swing production capacity compared with other sources of flexibility, especially interruption. In the UK market, approximately 26% of sales are on an interruptible basis, including 69% of sales to the electricity sector. Although the short-term fuel-switching capability of generators and industrial boilers is quite limited, the country's 16 GW of gas-fired power capacity can easily be interrupted without jeopardizing security of power supply, as the UK electricity generation industry has between 15 and 22 GW of over-capacity. This will obviously change as spare generating capacity is absorbed.

■ **An increasing swing in Norwegian production**

Although Norway was not considered a swing provider because of the high specific cost of offshore production and transport, the flexibility that Norwegian producers are willing to provide to their customers has increased during recent years, as seen in the graph below. The swing factor was 127% in 2000.

Figure 5 : Norwegian Monthly Gas Production (1993-2001)

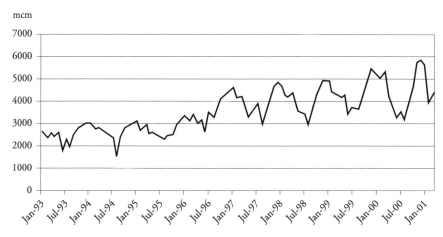

Source: IEA Monthly Database.

Gas from the Troll West field, a pure-gas field which accounts for about half of Norwegian sales, offers higher flexibility than the associated gas from Ekofisk. The first Norwegian gas sales in 1976 were from the Ekofisk area to the Continent and from the joint UK-Norwegian Frigg field to the United Kingdom. The sales contracts were based on the depletion of reserves in specified fields. Frigg deliveries had considerable flexibility because they were from a non-associated gas field, but Ekofisk had limited flexibility, as its gas was associated gas, whose exploitation was determined by oil-production needs. This situation changed after 1993 when deliveries to the Continent under the Troll Gas Sales Agreements began. The Troll contracts offer the customers more flexibility in annual volumes than the Ekofisk contracts. The increasing share of deliveries under the Troll Agreements is reflected in an increasing swing factor as shown in Figure 5. Under the Troll agreements demand is partly met by gas from smaller fields with Troll serving as a back-up when the smaller fields are shut in and to provide the contractual flexibility.

- **IEA North America**

North American gas production is mainly domestic. Only 1% of US supply is imported in the form of LNG from outside the region. 18% of US supply is imported from Canada with a relatively low swing. In the United States and Canada, gas production is steady throughout the year. Swing is 104% in the United States and 106% in Canada. These low figures reflect the economic constraint of long-distance transportation. Storage provides the major part of flexibility needs in the region.

- **IEA Pacific**

In the IEA Pacific region, most fields supplying gas to the region have the characteristics needed to provide flexible production. When the liquefaction chains were developed in the 1970s and 1980s, however, the high investment tied up in liquefaction plants and receiving terminals required even higher utilisation rates than for pipeline gas. The rigidity of LNG tanker schedules in the early days of LNG trade also limited flexibility. Since the investment in liquefaction plants is independent of distance, even short-haul LNG transport required high utilisation rates and so impeded flexibility of supply. In addition – at least in the beginning of LNG trade – there was little or no over-capacity in tankers, and tankers stuck to a well-defined schedule at both the loading and unloading points.

The swing factor of LNG imports in Japan is low at an average of 113% in 2000. Japan is still receiving LNG under long-term contracts signed in the 1970s and 1980s, with very little or no flexibility. The term of new LNG contracts will not reflect this rigidity of the pioneering days of LNG trade. To meet its high seasonal requirements, Korea buys spot cargoes during winter peak periods. Recently, spare capacity at new liquefaction plants in a number of Middle East and Asian countries has allowed more flexibility in LNG contracts and supplies. These new trends are analysed in Chapter 5.

Access to spare capacity at import and transportation facilities

Any additional supply that may be available from production and from flexibility in import contracts has to be transported to the place where this supply is needed. This requires that transportation capacity is available. Pipeline capacity is limited by the pipeline's diameter and maximum design pressure and by its compressor configuration – the distance between compressors and the ratio between the compressors' inlet and outlet pressure. Similarly, LNG transport capacity is limited at the receiving point by unloading capacity, tank capacity, regasification capacity and the availability of tankers. Apart from these physical restrictions, there may also be restrictions of access to capacity due to legal, contractual or regulatory restrictions.

In less mature LNG markets, spare capacity was not available to third parties. But in Europe, it is now available to third parties under a regulated or negotiated regime. The spare capacity in pipelines will be a factor to watch in the liberalised market. Bottlenecks in the transmission grid may make it impossible to transport available additional supply further downstream. With unbundling of the supply and transmission functions, a new risk is appearing that the physical infrastructure of the network will not always be sufficient to allow available gas to be flowed to markets.

STORAGE

Functions and characteristics

Withdrawals from storage facilities are a major source of temporary supplementary gas supply. For short-term requirements of from one to a hundred days duration, gas withdrawals from storage facilities often provide the quickest and most secure form of flexible gas supply.

Storage is a vital part of the gas chain. In the traditional markets, it performs three different functions to the gas operators :

- *Flexibility*
  - load balancing at any time, hourly, daily, weekly or seasonally, and more flexibility to end-users;
  - fulfillment of minimum take-or-pay obligations in times of low demand.
- *Security*
  Some European companies have built storage facilities as a buffer against interruption of supplies and they maintain strategic reserves to ensure security of supply.
- *More-efficient grid design*
  Storage allows a more-efficient design of the grid. Storage can cover peak demand and so the pipeline can be smaller and more fully used throughout the year.

With the introduction of competition, storage has new roles to play in gas trading. These new roles are analysed in Chapter 4.

Depending on the function, there are different requirements for the location of the storage. Storage intended mainly to provide flexibility is best located, if possible, close to the market it serves. Storage intended to make up for supply interruptions can be placed along the delivery route of the gas to be replaced, closer to the market than the sources of the most probable disruptions.

In traditional markets, storage facilities play a vital role in meeting peak demand. During the cold day of 2 January 1997, storage in France provided 1.3 TWh, or 52% of that day's gas supply – a rate sustainable for only a relatively short period.

Different types of storage are more or less suitable for the purposes described above. Suitability depends primarily on the volume and rates of withdrawal for which the storage facility is intended. Three major types of geological storage are used in IEA countries: depleted gas fields, aquifers and salt cavities. A few facilities are also developed in disused mines and a commercial project is under way to test storage in lined-rock caverns.

Depleted fields and aquifers can store a large working volume of gas at low cost. However, the withdrawal rate for a given working volume is limited. So they are best suited for seasonal balancing and for storing strategic reserves. Storage facilities built in salt cavities offer a high withdrawal rate for a given volume of working gas and can be cycled more than once a year. They are thus well-suited for daily or weekly balancing. However, their volume specific costs are higher than for depleted gas fields or aquifers.

Gas can also be stored in liquid form. However, because methane liquefies only at -162°C, storing gas in liquid form is very expensive. Storage in liquid form is therefore used only in the LNG supply chain at receiving terminals, or for "needle peak" management of remote parts of major transmission and distribution systems, in LNG peak shaving facilities.

Storage development in IEA

- **IEA North America**

In the United States, there are 415 underground gas storage facilities in 30 states. They have a working capacity of 110 bcm, representing 17% of average annual gas consumption in 2000. Canadian storage capacity is an estimated 17 bcm of working gas, or 19% of average annual gas consumption.

North American storage facilities were built primarily for commercial purposes to balance remote gas supplies and seasonal market demand, while minimising the size of pipelines and transportation costs. However, storage is increasingly being used for trading purposes. Several market hubs are now operating in North America, offering storage services such as gas loans, gas balancing and peaking services (see Chapter 4).

In North America, the continuing restructuring of the gas industry has resulted in increasing demand for new and more flexible storage sites, especially close to a trading hub. Storage capacity is expected to increase with rising demand for natural gas and with growing reliance on remote areas like Alaska.

■ **IEA Europe**

Europe has ample commercial storage capacity. The working capacity of the 94 facilities in IEA Europe is 58.8 bcm, or 13% of consumption. Storage capacity has grown rapidly in recent years. Since 1993, 13 new sites have been built in Western Europe, adding 15 bcm of capacity. Three countries dominate the European storage scene: Germany (with 39 storage sites and 18.6 bcm of working capacity), Italy (with 10 sites and 12.7 bcm) and France (with 15 sites and 10.5 bcm).

Future requirements for storage capacity will depend on two counteracting forces. On the one hand, the increasing share of supplies coming from more distant sources with high load will swell the need for new storage capacity. On the other hand, the increasing share of gas used as a base-load fuel in power plants will decrease the demand for swing and abate the need for new storage[21]. In addition, the opening of markets will provide more market mechanisms to make better use of spare capacity in pipelines and other infrastructure.

Europe has a large geological potential for all types of storage facilities. Planned enlargements of existing facilities and 30 new underground storage projects are expected to meet Europe's commercial storage requirements for the next 15 years. Furthermore, countries like the Slovak Republic and Latvia, which are close to the IEA European region, are also well endowed with gas storage capacity.

■ **IEA Pacific**

The use of underground gas storage is not common in the IEA Pacific region. Japan and Korea do not have underground gas storage, but rely on storage at their LNG regasification terminals.

---

21 See Chapter 2.

In Australia, the gas transmission network is not integrated because of the great distances between consuming areas. Four depleted gas fields have been converted into storage facilities to cope with seasonal demand fluctuations.

Liberalisation of the gas market is now under way in Japan and Korea. It may accelerate the expansion of gas transmission and storage. However, geographic conditions in Japan do not lend themselves to the building of transmission pipelines and underground gas storage.

Table 2: Underground Gas Storage in IEA Countries in 2000

Country/region	Number of storage facilities	Working gas volumes (million cm)	Peak daily deliverability rate (million cm/day)	Per cent of gas consumption (a)	Number of days of average consumption (b)	Number of days of firm demand (c)
IEA Europe	**94**	**58,761**	**1,243.98**	**13**	**47**	**116**
Austria	5	2,820	28.3	37	134	459
Belgium	2	654	11.3	4	15	42
Czech Republic	7	2,147	42.5	23	85	192
Denmark	2	810	24	17	60	311
France	15	10,490	182	26	95	182
Germany	39	18,556	425.2	21	75	157
Hungary	5	3,610	46.58	30	109	206
Italy	10	12,747	267	18	66	184
Netherlands	3	2,400	144	5	18	42
Spain	2	1,274	9.7	8	28	160
UK	4	3,253	63.4	3	12	26
IEA Pacific	**4**	**1,244**	**14.2**	**1**	**4**	**16**
Australia	4	1,244	14.2	6	20	106
IEA North America	**453**	**127,573**	**2912**	**17**	**64**	**177**
Canada	38	17,167	710.8	19	69	208
United States	415	110,406	2201.2	17	63	173
IEA	**551**	**187,578**	**4170.18**	**14**	**52**	**144**

(a) Ratio of working gas to annual consumption.
(b) Ratio of working gas to total consumption multiplied by 365.
(c) Ratio of working gas to residential and commercial consumption multiplied by 365.
Figures do not include peak shaving units.

Source: IEA, Natural Gas Information, 2002, and Country submissions.

Line-pack

Line-pack is the storing of gas inside the pipeline network by boosting the line pressure above the delivery pressure. It can be used as an initial means to balance supply and demand fluctuations during the day. Line-pack is a limited tool as it requires some time to build up, but it can be helpful especially when a cold weather front is coming.

Line-pack capacity depends on the design of the gas transmission system[22]. Some designs are based on the principle that transmission capacity and supply capacity are matched and the transmission system cannot be used for diurnal storage. In other designs, the transmission system can be used not only for transporting gas from the supply sources to the end-users but also as a means to balance the fluctuations in demand that occur during the day.

The use of line-pack differs strongly among countries, according to the design of the transmission grid and supply patterns. In the United Kingdom, for instance, line-pack has traditionally covered up to 3% of total demand. In Spain, the figure is 0.4%. In some circumstances, extra compression or a larger pipe may be a useful alternative to storage by increasing the potential for line-pack.

Storage at LNG terminals

Although limited, storage at LNG import terminals plays a key role in some importing countries, especially where geological options for underground gas storage are limited, such as Japan, Korea, Belgium and Spain.

In itself, LNG provides a valuable source of flexible supply and the growth in the short-term market for spot cargoes gives LNG buyers a chance to diversify imports still further and eventually to adapt supply to demand by using the LNG spot market. This issue is discussed in Chapter 5.

LNG terminals - except in the US - were not in the past open to third parties. So, storage capacity in LNG tanks was an operational tool under the control of traditional companies. This is changing in Europe, with the coming of third-party access to terminals, and the idea is being actively debated in Japan and Korea. This issue is discussed in Chapter 5.

---

22 GTE (2001).

Table 3: LNG Regasification Terminals in Operation in IEA Countries in 2000

	Number	Storage capacity (thousand cm of LNG)	Working capacity (million cm/day)
IEA Europe	**9**	**1,660**	**149.8**
Belgium	1	260	17.8
France	2	510	53
Greece	1	75	5.4
Italy	1	100	10
Spain	3	460	50.6
Turkey	1	255	13
IEA Pacific	**25**	**14,178.9**	**741**
Japan	23	12,178.9	604
Korea	2	2,000	137
IEA North America	**2**	**277.5**	**32.5**
United States (a)	2	277.5	32.5
IEA Total	**36**	**16,116.4**	**923.3**

(a) Everett, Massachusetts and Lake Charles, Louisiana.
The detail by country is given in Annexes.
Source: IEA (2002b) and Country submissions.

DEMAND SIDE MEASURES

Gas suppliers often conclude interruptible contracts with certain large industrial consumers and power generators. In return for a discount on the gas price, these customers reduce or stop their gas off-takes at the request of the supplier. The request is subject to certain criteria, often linked to temperature. "Interruptible" customers are usually equipped to switch from gas to other fuels or electricity, or to suspend the part of their production based on gas. Interruptible customers can make an important contribution to flexibility. Gas suppliers can arrange with these customers to decrease demand during times of tight gas supplies.

Interruptible contracts

Interruptible gas supply contracts enable the buyer to get gas at a discount of from 2% to 20% throughout the year[23]. In exchange, the supplier may

---

23 CERA (1998).

interrupt the gas flow according to mutually agreed criteria, normally depending on temperature. In some contracts, however, the supplier may interrupt deliveries as he sees fit, up to a maximum number of days per year. Customers, in turn, are free within limits defined by a minimum take-or-pay obligation - which may be zero on any single day - to vary the amount of gas they receive.

At present less than 15%[24] of gas sales in IEA Europe is on an interruptible basis. This is an average figure and individual country situations will be different.

Most customers with interruptible supply contracts have dual-firing capability. They are expected to hold non-gas capacity and fuel in reserve, so that they can continue to operate during an interruption. Nevertheless, Continental European customers have rarely been interrupted as often as the contractual terms allow and there is reason to question whether they are really interruptible. This could change in a more competitive environment.

In the US and the UK, interruptible customers are really interrupted and are therefore really interruptible. Interruptible contracts are not just a way to offer discounts to a certain class of customers, as is the case in some Continental European countries, but a management tool for balancing supply and demand. In the United States, in 1998, total sales on an interruptible basis amounted to 140 bcm or 25% of total gas sales. Thirty-eight per cent of sales to the industrial sector were on an interruptible basis (93 bcm), 36% of sales to electric utilities (33 bcm) and 15% of sales to commercial consumers (13 bcm).

In the United Kingdom, sales of gas on an interruptible basis accounted for 26% of all gas sales in 2000, mainly to electricity generators. Interruptible sales to the electricity generation sector were an estimated 68% of total sales to that sector. Due to over-capacity in the UK electricity industry, interruptions of gas-fired power plants have no effect on electricity sales.

Fuel-switching

For decades, natural gas has competed with fuel oil in industrial markets. Depending on the use, it may be possible to switch from gas to another fuel without any interruption in the production process. But in uses such as making sheet glass, where the gas flame is directly applied to the material, it is very hard to switch at short notice to another fuel.

---

24   Most gas supplying companies hold a substantial share of interruptible contracts with customers they never interrupt in practice. As a result, effective short-term switching capacity may be lower than usually stated.

The gas use most suitable for fuel-switching is in boilers, either for steam raising or for power generation. For steam raising purposes in the industrial sector, gas can be replaced by residual fuel oil. In the electricity sector, with more and more gas used in combined-cycle gas turbine plants, the only alternate fuel will be distillate. Exact figures on various countries' real (as opposed to nominal) capacity for rapid switching from gas to other fuels are hard to find. A recent IEA survey[25] identifies short-term fuel-switching capability in IEA countries. The survey focuses on short-term fuel-switching from oil to other fuels in the context of an emergency response measure. But it also examines the effect that gas supply disruptions might have on oil demand and, in this context, provides information on the short-term fuel-switching capability of the power and industrial sectors in IEA countries.

Figure 6 below illustrates the results of the study. For the IEA as a whole, fuel-switching capability from gas to oil by industrial customers and power generators is estimated at 3.5 mb/d (corresponding to about 490 mcm/d). This result does not take into account the utilisation rates of plants. It represents a maximum rated capability.

This 490 mcm/d is concentrated in only five countries: the United States (165 mcm/d), Japan (106 mcm/d), Korea (48 mcm/d), Germany (25 mcm/d) and Italy (25 mcm/d).

Figure 6: IEA Fuel-Switching Capabilities (1999 data)

Total IEA fuel-switching from gas into oil 490 mcm/d (3.5 mb/d)

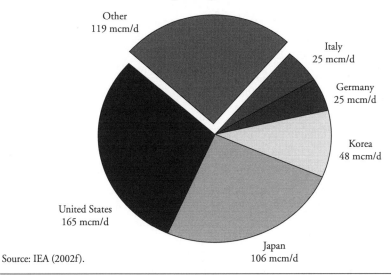

Source: IEA (2002f).

25 IEA (2002f).

These results suggest that name-plate short-term fuel-switching capability by industrial customers and power generators amounts to about 12% of average daily gas consumption in IEA Europe, to 9% for IEA North America and to 50% for IEA Pacific. The figures should be considered as maxima, since the switching capabilities indicated might no longer be effective, either because local sources of residual fuel oil in some countries no longer exist, or because the equipment has not been maintained. In some countries, new environmental constraints rule out switching to a more polluting fuel.

In principle, all gas-fired power generation could be cut off and replaced by electricity fed into the electric grid so long as there was enough spare non-gas-generated power and transportation capacity. This could be done simply to bring costs down but it could also be done to compensate for interruptions in gas supply.

Additional information on fuel-switching capacity in the power sector is available in Table 4. The table shows the share of thermal generating capacity

Table 4: Multi-Fired Electricity Generating Capacity in IEA Countries at 31 December 2000 (GW)[26]

	IEA North America	IEA Europe	IEA Pacific
Total electricity generating capacity (public utilities + autoproducers)	944.16	682.22	365.26
Total capacity by combustible fuels	663.19	350.24	242.80
Of which Single fuel-fired	394.64	220.84	206.05
Of which Multi-fired	268.55	129.40	36.75
Solid-Gas	40.29	8.72	1.00
Solid-Liquids	6.29	47.59	4.62
Liquids-Gas	56.98[27]	55.72	30.49
Liquids-Solids-Gas	164.99	17.37	0.64
% multi-fired/total generating capacity	28%	19%	10%
% multi-fired/thermal generating capacity	40%	37%	15%
% liquids-gas/multi-fired generating capacity	21%	43%	83%
% liquids-gas/total generating capacity	6%	8%	8%

Source: IEA (2002g).

26 The figures in the table are maxima, as the rate of utilisation is not taken into account.
27 ESAI, an independent research firm based in Boston, estimates at about 60 GW the true dual-fired gas /heavy fuel oil capacity in the US (ESAI, 2001).

in total electricity capacity in the three IEA regions, the share of multi-fired power plants and the plants that can switch from or to natural gas.

There are important differences by region. IEA North America has a large multi-fired capacity, 28% of total electricity generating capacity, compared with 19% for IEA Europe and 10% for IEA Pacific. The share of multi-fired power plants in thermal generating capacity is also higher in IEA North America (40%) than in IEA Europe (37%) and IEA Pacific (15%).

Natural gas plays a large role in switching capacities. Twenty-one per cent of multi-fired power plants[28] in IEA North America can switch from oil to gas and back again. This represents 6% of total electricity generating capacity and can exert a great effect on markets and prices, in particular in very cold weather. This is further analysed in Chapter 4. Forty-three per cent of multi-fired power plants in IEA Europe can switch between oil and gas. This represents 8% of total capacity. In IEA Pacific, 83% of multi-fired power plants, 8% of total capacity, can alternate oil and gas.

Demand-side reaction to price signals

Gas customers with fuel-switching capabilities, whether industrial customers or power generators, may themselves choose to switch fuels in response to market signals. In Europe, many power generators can call on alternative generating facilities that use other fuels. Increasingly, they may call on electricity provided by the power market. But, as a matter of fact, industrial fuel-switching in response to price signals is limited in Europe, due to the practice of pegging the gas price to the price of substitute fuels.

In the United States, on the other hand, many industrial consumers have the option of switching to other fuels when natural gas prices rise[29]. This option was exercised by some industrial users in 2000-2001, when natural gas prices rose dramatically. Another option chosen by some industrial users was to reduce or stop operations and sell off gas they had already contracted and paid for to the highest bidder. Examples include Terra Nitrogen, which shut down one of its fertiliser plants in 2000 and cut back operations at another, and Mississippi Chemical, which halted fertiliser production altogether. Both companies sold off their natural gas futures contracts[30].

When spot prices in the New York City market rose from $2.65 to $15.34/MBtu in January 2001, there was evidence of fuel-switching by gas customers as well

---

28 If only liquid/gas power plants are taken into account.
29 According to EIA (1997b), 39% of industrial natural gas consumption in 1994 could have switched to other fuels. The IEA Survey (IEA 2002f) indicates a rate of 28% in 1999.
30 EIA (2001b).

as interruption of gas service to non-firm customers in the Northeast US. The steep rise was due to a sudden surge of cold weather and a deliverability constraint.

The responsiveness of demand ultimately depends on how much the response costs, relative to the prevailing natural gas price, and the time required to implement that response. The moderate responses to very high US gas prices in late 2000 and early 2001 in US suggests that the costs of fuel-switching could be higher and the delays longer than might have been thought[31].

Table 5: Flexibility in IEA Countries in 2000[32]

	IEA North America	IEA Europe	IEA Pacific
Production swing	105%	134%	116%
Import swing	119%	118%	113%
Supply (production + import) swing	107%	125%	109%
Consumption swing	137%	152%	111%
Storage (working gas as a percentage of annual consumption)	17%	13%	1%
LNG tanks (a) (tank capacity as a percentage of annual consumption)	0.02%	0.2%	7%
Fuel-switching as a percentage of average daily gas consumption (b)	9%	12%	50%
Share of temperature-sensitive consumers in total consumption	35%	40%	23%

(a) 1 cm of LNG = 593 cm of gas.
(b) Calculated from the IEA survey on fuel-switching (IEA 2002f). This daily figure cannot be translated into an annual figure, as fuel-switching capability and interruptions are normally limited to a number of days.

31 EIA (2001e).
32 The regional aggregated figures in this table are representative for IEA North America, which can be considered as one market. The situation is different in IEA Europe and Pacific, where the situation of individual countries can be quite different from the average one. The lack of interconnections with sufficient capacity may restrict flexibility provisions in some peripheral European countries.

CONCLUSION ON THE CURRENT STATE OF FLEXIBILITY

At present storage volumes, the number of interruptible supply contracts, swing potential in indigenous production and import contracts in IEA gas markets are all high, as indicated in Table 5, though the situation in individual countries may be quite different. Circumstances will evolve with the continued growth in gas demand, new supply trends and market liberalisation. The impact of these new trends is discussed in Chapters 4 and 5.

4

MARKET LIBERALISATION: A NEW CONTEXT FOR FLEXIBILITY

INTRODUCTION

Until now flexibility in traditional gas markets has mainly been a matter of volume, and has been provided by traditional instruments, such as supply swing, interruptible contracts and storage. The role of storage was confined largely to meeting variations in demand and interruptions of supply. There were few short-term opportunities to arbitrage between different energy commodities and different markets. As markets are being opened through third-party access, as well as by abolition of state monopolies and exclusive concessions for transport and distribution, competitive markets for gas are emerging and new gas services are being developed. Gas flexibility in its various forms is becoming a tradeable service and is valued by the market. Experience in North America and the United Kingdom indicates that TPA leads to gas-to-gas competition and the emergence of marketplaces, usually trading hubs. Gas is traded like any other commodity on such a marketplace. Once there is sufficient liquidity, spot markets for immediate and forward delivery emerge. In places with a liquid forward market, futures markets evolve to hedge exposure to price volatility. Price volatility modifies the role played by traditional flexible tools, such as storage, which is increasingly used or price optimisation rather than simply to manage volume.

With the opening of gas and electricity markets, opportunities for price arbitrage between the two commodities arise. In countries where gas is used for power generation, arbitraging itself permits a completely new instrument to provide flexibility. With arbitrage between the gas and electricity markets, gas demand and supply can be partly brought in line by drawing on supply and demand in the electricity sector.

This chapter analyses the new flexibility emerging with market liberalisation. The first section deals with the emergence and development of gas trading, with reference to the North American and British experiences. It also discusses the latest trends in Continental Europe, in particular the recent emergence of trading hubs for gas. The second section analyses the emergence of flexibility trading and the changing role of traditional flexibility instruments in this new environment, especially that of underground gas storage. The third section discusses the new features and opportunities arising from the development of gas/power arbitrage that results from the development of liberalised markets for natural gas and power.

THE EMERGENCE AND DEVELOPMENT OF GAS TRADING

Marketplaces: new hubs

With the introduction of third-party access, new trading opportunities are emerging in places where different owners' pipelines meet each other, as one owner may now use the other's line[33]. Some places, where several pipelines meet and where storage facilities and consumption centres are both close by, develop into a marketplace for gas – a hub. Hubs provide the means to increase short-term exchanges between parties; short-term/short-haul transportation services; and opportunities to reduce price risk exposure[34].

For a hub to qualify as a market, the hub operator must provide the following services: interconnections between the pipelines to allow the gas to be interchanged between the systems and, ideally, storage facilities; ease of transportation to and from the hub; and several associated services, such as balancing[35] and recording of title transfers.

An important prerequisite for successful short-term trading at a hub is the speed at which contracts can be concluded. Standardised contracts expedite trading. An active approach on the part of participating gas operators can also contribute to the success of a hub.

Short-term balancing of supply and demand is a major function of a market hub. Storage at hubs, particularly when they are close to markets, is a valuable tool for traders. Storage adds flexibility to the marketplace, both in physical terms, via access to gas (at times of peak demand) and in trading terms, by providing physical hedging and the ability to arbitrage.

The number of sellers and buyers at a hub is important: it needs to be large enough to create a liquid market, where there is sufficient supply and demand for gas to be traded rapidly and freely. It also needs to be large enough so that one transaction will not alter the market price. The size of the "churn" – the

---

33 CEC (2000).
34 EIA (1997a).
35 A short-term interruptible arrangement to cover a temporary imbalance between supply and demand.

ratio of traded volumes at a hub to actual physical volumes – is generally deemed the yardstick for when a hub becomes a successful pricing reference.

In liberalised gas markets, two different types of marketplaces have emerged. On the one hand, there are gas hubs, like Henry Hub in the United States or the Alberta Energy Company (AECO) hub in Canada. On the other hand, there is the national balancing point (NBP) in the UK.

The NBP is a virtual trading place covering the whole network of Transco, the UK gas transmission company. Due to the short average transport distance and the high number of inlet and outlet points at relatively short distance, the UK system is like a grid. Gas on which the entry fee has been paid is treated as being at the NBP, in other words, on the marketplace. To take the gas from the NBP an exit fee has to be paid. As on traditional marketplaces, transport costs to and from the marketplace (entry and exit fees) may differ for individual market participants but this is irrelevant to the trading of the commodity on the marketplace.

In 1994, the NBP became an informal market for gas trading among UK power generators. Transco's introduction in September 1996 of a tight daily balancing system, backed by penalties, led to NBP's using geographical locations – such as Bacton, the main landing point for North Sea gas in England, or its Scottish equivalent, St. Fergus – as the main places for spot gas trading activity. Approximately 15% of the gas at the beach[36] is sold directly to the wholesale spot market. A similar proportion is sold as one-to-three year contracts, with the remainder being sold on traditional long-term contracts, most of them indexed to oil prices. Currently, about 80 traders are active on the NBP and the churn ratio is 17 to 1.

In the US, the 278,000-mile gas pipeline system includes numerous pipeline interconnections. Before the establishment of open access in 1986, little could be gained from using these interconnections. Open access to transport and storage has completely modified the picture. It is now possible for everyone to move gas between pipeline systems and between pipeline and storage systems. And this has engendered hubs as natural transfer and trading points of gas.

North America has 39 trading hubs, of which Henry Hub is by far the largest. In the United States, gas companies promoted hubs to increase trade in gas and capacity across pipeline and storage systems and to meet the need for short-

---

36 In the UK, when gas has been brought ashore to a terminal by producers but is not yet in the National Transmission System, the gas is called "at the beach".

term balancing services that were formerly provided by pipeline companies under bundled services[37].

Henry Hub in southern Louisiana is the largest hub in the world. It connects twelve pipelines and has access to three salt storage caverns. It is accessible to major producers from both onshore and offshore Louisiana. Information on prices and other relevant matters is readily available. It has a very high liquidity: its churn ratio is approximately a hundred to one. Henry Hub also serves as the delivery and reference point for the New York Mercantile Exchange (NYMEX) gas futures contract and is the reference for all gas export contracts to Mexico.

Box 3: US Market Centres and Hub Services

The type of services provided by market centres and hubs varies significantly. The Federal Energy Regulatory Commission's Office of Economic Policy provides a comprehensive list of the available services. The major ones are:

Balancing – A short-term interruptible arrangement to cover a temporary imbalance. This service is often provided in conjunction with "parking" and "loaning".

Electronic Trading – Trading systems that either electronically match buyers with sellers or facilitate direct negotiation for legally-binding transactions. Customers may connect with the hub electronically to enter gas nominations, examine their account position, or access e-mail and bulletin-board services.

Loaning – A short-term advance of gas to a shipper by a market centre that is repaid in kind by the shipper a short time later. Also referred to as "advancing", "drafting", "reverse packing" and "imbalance resolution".

Parking – A short-term transaction in which the hub holds the shipper"s gas for redelivery at a later date. Often uses storage facilities, but may also use displacement or variations in line-pack.

Peaking – Short-term, usually less than a day and sometimes hourly, sales of gas to meet unanticipated increases in demand or shortages of gas experienced by the buyer.

Storage – Storage that is longer than "parking", such as seasonal storage. Injection and withdrawal may be separately charged.

---

37 EIA (1997a).

Title Transfer – A service in which changes in ownership of a specific gas package are recorded by the market centre. Some gas titles may be transferred several times before the gas leaves the centre. This service is merely an accounting or documentation of title transfers. It may be done electronically and/or by hard copy.

Wheeling – Essentially a transportation service. Transfer of gas from one interconnected pipeline to another through a hub, by displacement, including swaps, or by physical transfer over a market-centre pipeline.

Source: EIA (1997a).

Spot and futures markets

Once trading at a hub develops into a liquid market, spot and futures markets will form and a market price for immediate and future delivery will emerge. Spot markets usually start with over-the-counter trades based on standardised agreements for a fixed volume of gas. They are made either bilaterally between the two parties concerned or through a broker. Gas delivery can be for periods of between one day and one year, either for prompt (very short-term) or forward (long-term) delivery at a defined location, usually the hub.

Deliveries in the future are dealt with in forward contracts, which are a commitment to deliver or take a specific amount of gas, usually in units of 10,000 MBtu, at a defined time and place for an agreed price. The financial transaction takes place on the day of delivery. Forward contracts are traded over the counter, in customised one-off transactions between a buyer and a seller.

Futures markets emerge in countries that have fully liquid spot markets for immediate and forward delivery. Although they have similar names, "forward" contracts and "futures" contracts are quite different instruments. Gas futures are usually paper trades that track the daily movement of the expected future price until the expiration date of the contract, when gas must be delivered or the differential between the agreed price and the spot price on that day must be settled in cash. Unlike forward deals, which may be traded over the counter and always related to final physical delivery, futures contracts are traded on organised commodity exchanges with standardised terms. Futures contracts – because they are financial hedging instruments – can be traded independently from delivery to the underlying spot gas marketplace. They nevertheless need a spot market as a final reference point.

Futures markets provide an independent and transparent pricing signal for future price development and this can be used as a pricing indicator for other contracts.

The future price represents the current market opinion of what the gas will be worth at some time in the future. It is the only indicator (although by no means a correct prediction) of the expected spot price of the commodity in the future. Futures prices also serve as a stimulus to store or release gas from storage.

The other main function of the futures market is to transfer risk. Hedging allows market participants to lock in prices and margins in advance. Hedging reduces exposure to price risk by shifting it to those with opposite risk profiles or to speculators who are willing to accept the risk in exchange for possible profit. By using futures contracts, anybody who is dependent on gas prices may offset or minimise the risk inherent in a fluctuating gas price. A gas buyer may have an interest in buying gas futures within a certain price limit to hedge the cost of using gas as an input into his productive activity. A seller, perhaps a small independent producing company, may want to hedge its earnings to meet its minimum income requirements such as interest payments on its financing.

Three broad categories of traders can be identified in futures markets – hedgers, speculators and arbitrageurs. Hedgers enter the market to offset a position with the intent of managing risk. In hedging one transaction is protected by another. Speculators are willing to accept the risk in exchange for profit. They take a position with the intent to earn a margin. Speculators provide the market with liquidity. Arbitrageurs take advantage of momentary disparities between prices of the commodity in two different marketplaces. They make markets more efficient by bringing prices of different marketplaces in line with each other. Increasingly, arbitrageurs will trade across commodities with the effect that pressures on gas prices may have their origins in non-energy commodities.

Various types of regulations have been introduced to prevent undue volatility in or influence on the markets. These regulations include limits on daily price movement, position limits and surveillance to detect trading irregularities.

NYMEX launched the world's first natural-gas futures contract in April 1990 with Henry Hub as the reference point. Volumes and "open interest" (the number of futures or options natural gas contracts outstanding in the market) have grown rapidly, and the NYMEX gas contract is the fastest growing instrument in the exchange's history. The estimated trading volume, around 725 Btu (20 bcm) of gas a day, is ten times the amount of gas delivered daily in the United States. In October 1992, NYMEX marked another milestone in the energy markets when it launched "options" on natural-gas futures, giving market participants still another instrument to manage their market risk.

The exchange allows hedgers and investors to trade anonymously through futures brokers. NYMEX gas instruments have attracted private and institutional investors who seek to profit by assuming the risks that the industry seeks to avoid in exchange for the possibility of rewards. A wide cross-section of the gas industry is active on the futures market from producers to end-users: producers, processors, local distribution companies and marketers, industrial and commercial gas users. The marketers, predictably, are the largest participants accounting for 69% of the open interest in 2000.

Gas suppliers use the NYMEX futures contracts to provide a variety of services to help customers manage their price risks. These include fixing gas costs or expenditures and offering ceiling prices. Marketers often manage the market risk for their customers. As a result, local distribution utilities themselves accounted for only 1.7% of the reportable open interest for 2000 – although a much larger share of transactions was performed on their behalf.

In Canada, the "Natural Gas Exchange" (NGX), located in Calgary, provides electronic trading and clearing services to natural-gas buyers and sellers in Canadian markets. The NGX started service in February 1994. Over the past eight years, NGX has grown to serve over 150 customers with trading activity averaging 200,000 TJ (5 bcm) per month. NGX is wholly owned by the OM Group, the world"s leading provider of transaction technology. In Canada, the regulation of commodity trading is under provincial jurisdiction. The Alberta Securities Commission is NGX's lead regulator.

In the United Kingdom, the International Petroleum Exchange (IPE) launched gas-futures contracts in 1997. They are based on deliveries of natural gas at the NBP. The IPE gas futures market is a transparent, screen-based system, which provides a mechanism for risk management, hedging and in some cases the physical delivery of gas. The IPE traded a daily average of 60 million therms of natural gas, or approximately 60% of the UK's daily consumption.

Price volatility[38] is inevitable in competitive markets. It is a fact of economic life in all commodity markets. When the industry operates close to full capacity, small changes in supply and/or demand or relevant news items or sound bites may cause strong market pressures and substantial price increases or decreases[39]. This was illustrated in the United States in late 2000, when gas supply constraints led to a price surge.

Excessive and sustained price fluctuations can put large, capital-intensive supply projects at risk, especially those that require long lead times, such as LNG terminals and long-distance pipelines. In the US, sustained large shifts in price may preclude the building of new LNG facilities even though they could

---

38 Volatility is measured as the relative deviations around an average price value.
39 EIA (2001e).

Figure 7: Spot and Forward Prices in UK and US

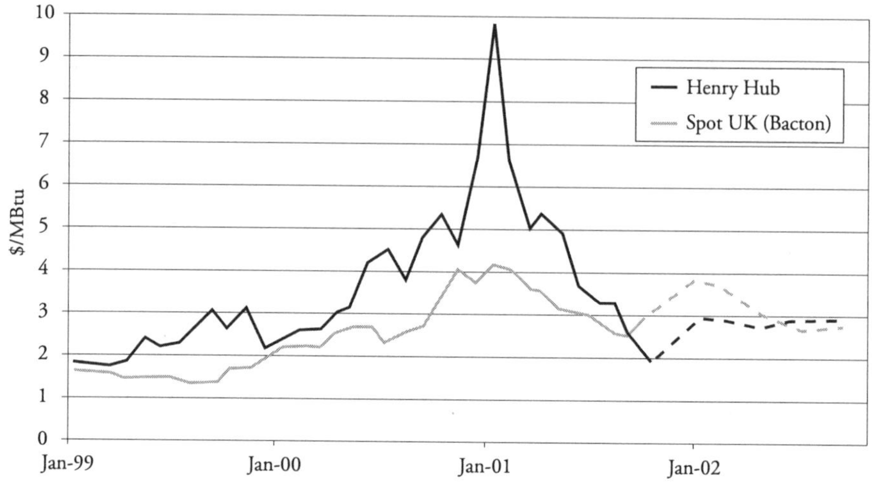

Source: World Gas Intelligence, European Gas Matters.

moderate natural-gas price fluctuations in the future. This problem is common to all volatile energy sources, notably in the oil markets, where the oil price collapse in 1998/99, followed by high peaks in 2000, created major problems for investors.

It is hard to identify and weigh causes of price volatility. However, it is essential that the new markets work efficiently, free of the abuse of market power by well-positioned players and of inappropriate trading practices. For instance, although the market needs speculators to ensure liquidity, their actions can seriously affect the price of the commodity. During 2000, when US gas prices soared, it appears that a huge inflow of speculative money exacerbated upward pressure on prices. Gas price trends in that period clearly reflected overreaction to the actual imbalances in supply and demand. Trading can thus cause additional volatility – volatility not based on responses to perceived changes in current or future fundamentals but on the desire for pure trading profits.

The US exchanges are under the control of the US Securities and Exchange Commission (SEC). With the emergence of increasingly complicated risk-hedging instruments, there is a rising risk of misuse or lack of control, as was demonstrated by the tremendous pressure on energy merchants in the wake of the spectacular bankruptcy of Enron in 2002. However, even this failure did not disrupt the working of the gas markets, which were liquid enough to bridge any shortfall that might have been caused by Enron's disappearance from the trading floor. As a consequence of the Enron bankruptcy, the stocks of some major US energy companies have fallen more than the stock averages over the past twelve months. Some of them are also under investigation by the SEC and the Federal Energy Regulatory Commission for possible trading or accounting irregularities. Serious questions surround the issue of risk-hedging operations and overly "creative" accounting practices.

The recent events in US energy markets have brought to the fore the need for mitigating counter-party risk in gas trading and ensuring that trading markets do work fairly and efficiently. Spot and futures markets do offer new flexibility to individual buyers and sellers. But if the system does not send the proper market signals so that the underlying physical flexibility instruments are built in time to cover variations in supply and demand, the market will remain very volatile. Volatility is unlikely on over-supplied markets, but is a real threat when the market is tight.

Emerging trends in Continental Europe

■ Zeebrugge Hub

The Zeebrugge hub in Belgium is the first gas-trading hub to be launched in Continental Europe. It is located at a coastal town where gas pipelines from the United Kingdom (Interconnector) and Norway (Zeepipe) meet. There is also an LNG import terminal and a link into the Belgian national gas transmission network. Zeebrugge is linked by large pipelines to France, the Netherlands and Germany. These pipelines and terminals are all interlinked, so that gas can physically be moved or exchanged between them.

The Belgian company Distrigas with the support of 40 other gas companies launched Zeebrugge as a market hub in November 1999. It is operated by a Distrigas subsidiary, Huberator. Huberator has two functions: to manage physical gas flows between the different inlet and outlet points in Zeebrugge and to act as a broker between the partners using the Zeebrugge hub. Three main services are offered: matching nominations[40] (verification and confirmation of corresponding trade nominations), title tracking and allocation.

Distrigas has developed two standard contracts: a Hub Services Agreement (HSA) between the hub operator and parties using the hub, in essence a rulebook on hub operations, and the Zeebrugge Gas Trading Terms and Conditions, under which two parties may enter into a standardised gas sales agreement at the hub.

---

40   The notification to put into effect a contract or part of a contract, e.g., a gas flow nomination from a shipper to advise the pipeline owner of the amount of gas it wishes to transport or hold in storage on a given day.

Under the HSA, Huberator undertakes to provide dispatching, balance checks and matching services on a 24-hour basis. Balance checks are necessary, because no shipper is permitted to have a gas surplus or shortfall at the hub at the end of the day. Huberator also confirms effective hourly deliveries, redeliveries and allocation – when capacity constraints on pipelines leading to the hub require a prorationing of shippers' nominations.

The Zeebrugge trading document is a physical gas contract, priced in euros per gigajoule under which volumes in GJ have to be matched each hour. Although the trading document is voluntary, it has gradually been adopted for all Zeebrugge transactions. Payments due in the event of mismatched deals are settled by the traders involved, without involvement by Huberator.

The churn ratio at Zeebrugge is seven-to-one. Forty-four companies now actively trade at the hub. Two elements have been of critical importance to the successful development of the hub. One was the opening of the UK-Belgium Interconnector in October 1998. This enabled arbitrage between the UK spot market and the Continental market where virtually all gas business was still done on a long-term basis, with gas prices indexed to oil. The other was Distrigas's keen interest in developing Zeebrugge as a commercial gas hub.

■ Bunde

A second hub is being developed at Bunde in Germany. Bunde is a crossing point for three important pipelines. The first carries Dutch gas from the delivery point at Oude Stadenzijl to the east and south of Germany. The second carries Norwegian gas from the Emden/Dornum landing points to the South of Germany. The third, the Midal system built to compete with the existing pipeline system, links North Sea gas to Russian gas imports. The Etzel, Dornum and Rheden storage facilities are close by and Bunde is also close to the Ruhr area, a major industrial area of Germany.

Two groups of companies are preparing to launch two parallel hubs. One is being developed by EuroHub BV, which is wholly owned by Gas Transport Services, the transport arm of the Dutch company Gasunie. The other is the work of NWE-Hubco, a joint venture between the two German companies, Ruhrgas Transport and BEB Transport, and the Norwegian company Statoil.

Gas Transport Services is concentrating on offering capacity at the pipeline juncture and title tracking. Since the end of February 2002, EuroHub has provided a title-transfer service in the high-calorific Oude- Statenzijl-Bunde-

Emden gas network. It is planning to expand its hub services soon to potential further hubs in the Netherlands.

Ruhrgas, BEB and Statoil came together in November 2001 to form the North West European Hub Company to manage the logistics and infrastructure of building a new marketplace. NWE-Hubco intends to offer the following services:

- Safeguard the physical hub balance (net flow).

- Provide and manage information technology infrastructure for an integrated commodity market.

- Provide and manage IT infrastructure for a secondary market for transportation capacity and wheeling services – the transfer of gas from one interconnected pipeline to another through the hub.

- Provide and manage IT infrastructure for wheeling services between the pipeline junctures at Oude/Bunde or Emden.

- Nominate, allocate and track titles for gas traded at the hub.

A drafting group has been established to prepare standard trading terms and conditions for both Eurohub and NWE-Hubco.

This move in northern Europe is extremely important for future trading. Given its favourable location, Bunde could well develop into a true European hub.

- **Hubs under discussion**

With market liberalisation spreading across Europe, other trading hubs are being discussed at strategic points on the European network, where several pipelines interconnect and there is proximity to storage facilities and demand centers:

- Baumgarten in Austria,

- Lab in Slovakia,

- Lampertheim near Ludwigshafen, in Germany,

- One hub in southwest of France or Spain,

- The Po valley in Italy.

Figure 8: Existing and Possible Gas Hubs in Europe

NEW FORMS OF FLEXIBILITY AND THE CHANGING ROLE OF GAS STORAGE

Unbundled services

Third-party access enables buyers to choose their supplier, and this frees them from dependence on an all-in-one service. In effect, they can buy gas *à la carte* instead of from a standard menu. The opportunity has opened up for trading not only gas but also all kinds of services. The Dutch company Gasunie was the first of the Continental gas companies to offer unbundled services *à la carte*, such as peak capacity, seasonal swing and temperature-related backup. Most other gas companies have followed suit and are now offering a similar set of unbundled services.

Box 4: Gasunie Flexibility Services

Gasunie Trade and Supply[41] (GTS) is one of the largest providers of services associated with the supply of natural gas, such as capacity, flexibility and back-up. These services can be booked separately and the user is not obliged to buy gas from Gasunie.

In the case of additional capacity requirements, however, the customer must reserve corresponding transportation capacity. Customers who exceed the capacity reserved in their contracts are not cut off, but they must pay a penalty for the additional capacity required.

Capacity is based on hourly rates.

Flexibility services offered:

■ Additional Annual Flexibility

In addition to the flexibility that comes with annual off-takes provided by Gasunie as a non-exclusive supplier, Gasunie offers additional flexibility in annual off-takes as a separate flexibility service.

■ Additional Capacity

In addition to the base-load capacity included in gas-supply contracts with Gasunie or another supplier, Gasunie offers additional capacity on an annual basis. It may be used throughout the year along with the capacity already booked from Gasunie or another supplier. This additional capacity would not include additional volumes. Typically it would be used to cover seasonal variations in demand.

■ Incidental Capacity

Incidental capacity is offered to cover incidental peaks in the off-take pattern. Gasunie offers this service to customers whose maximum hourly capacities are not often required. There is a maximum of 31 periods a year of 24 hours each.

■ Hourly flexibility

Hourly flexibility gives the customer the right to receive for a limited number of consecutive hours, a certain capacity in addition to its contractual and possibly its incidental capacities. Hourly flexibility is a possibility where there is a cyclical pattern to the customer's use – variations between day and night. In hourly flexibility, a customer contracts for a "virtual" buffer that fills up during hours when demand is lower than the contract provides for, and empties when demand is higher.

---

41 The organisational division of N.V. Nederlandse Gasunie was implemented on 1 January 2002. There are now two units: Gasunie Trade and Supply, a trading company buying and selling natural gas and gas-related services in the Netherlands and elsewhere in Europe, and Gas Transport Services, for transport and related services. This is an interim step toward the planned legal division of the company.

The unbundling of services has had several consequences for gas customers and for the companies' operations. In a liberalised market, eligible customers – gas users that meet criteria specified in the EU Gas Directive or in national legislation, such as a minimum volume of gas consumed per year, have the right to choose their supplier and request third-party access to the grid – are free to decide for themselves how much flexibility and back-up they require, and will contract this from their gas suppliers or other service providers. Responsibility for procuring adequate flexibility lies with the customer. The various commercial players merely fulfil their contractual obligations. They do not take responsibility for ensuring the overall flexibility or security-of-supply needs of their eligible customers. A new risk pattern has emerged. Instead of contracting for a guaranteed supply under all circumstances, including temperature variations, each eligible customer can design for himself the risk pattern that suits him best and pay accordingly. But eligible customers have to bear the consequences of not contracting adequate services. They may be cut off or may have to pay penalties for exceeding their agreed-upon draw on the system.

Providing eligible consumers with the choice to buy tailor-made flexibility leads to better use of existing infrastructure. It also sends out market signals and creates a transparent market value for flexibility-related services. It allows the customer to choose between using the services offered by the system, or any alternative flexibility services belonging to him, or any kind of financial risk-hedging available on the market. The customer may also sell gas and services, bought but not needed by him. The allocation of scarce capacity is decided by market mechanism.

The question remains, however, of how non-interruptible customers, who have no fuel-switching capability, will get secure supply including the flexibility needed to cover exogenous influences.

The new functions of underground gas storage

As gas market structures change, storage offers new business opportunities, mainly as a trading tool to ensure price optimisation. In the United States, gas storage has become an independent service, both physically and in the context of financial trading. Traditionally, storage facilities were owned by local distribution companies and inter- and intra-state pipelines. Mega-marketers have bought a lot of available storage, since it allows them to take best advantage of arbitrage opportunities and price swings. Since storage facilities were unbundled from transportation, additional services have evolved, such as parking, balancing and loaning. More recently storage operators have combined the physical services of storage with financial derivatives. More than two-thirds of market centres in the US have access to storage.

■   Arbitrage on price variations and trading support

Storage now allows a trader to exploit price differentials between different points in time. One can, for instance, store gas in summer when prices are low and sell it in peak winter time, when prices are high. This is known as the "seasonal spread".

Market liberalisation has brought price volatility, and storage is a useful tool to benefit from price variations. One can buy gas cheap and store it with the intention to sell when prices go up. Price arbitrage between spot and futures gas prices is profitable, if the cost of storage is less than the price differential between the present and the future. The difference between the current spot price and the futures price for nearby delivery contracts in future months is called the "flex spread". When supplies of gas are tight, the spot price can move above the futures price. In this case, a company can sell gas from storage onto wholesale markets and, at the same time, buy more gas under a futures contract. The company is guaranteed the difference in these prices less any transaction costs.

This approach calls for much more operational flexibility than did a traditional seasonal supply service. The operator must be able to inject or withdraw gas at any time during the year at short notice.

■   Impact on spot prices

A study carried out by the US Energy Information Administration[42] indicates that expected storage requirements and spot prices are strongly linked during the winter heating season. High prices in the spot market are generally accompanied by low levels of storage relative to expected deliveries. As the temperature drops below normal, more and more gas is withdrawn from storage and prices can rise dramatically. During the rest of the year, the relationship between storage and spot prices is much less direct.

In the United States, the amount of storage anticipated throughout the year is a key element in gas prices. Storage refills can determine the near-term direction of gas prices. The gas supply situation during the spring and summer bears close monitoring. If it is particularly hot in regions that consume large quantities of gas-fired electricity, then injections into storage for the next winter drop off, and this brings sharply rising prices during the so-called "injection season".

---

42   EIA (1995).

Box 5: Impact of Storage Levels on Gas Prices: Recent US Experience

In the United States, spot gas prices at Henry Hub began to climb during the first half of 2000 and reached record highs, exceeding $10/MBtu at the end of December 2000. The principal factors behind the surge were the supply-and-demand fundamentals. This was exacerbated by low storage. On 1 November 2000, at the beginning of the heating season, total working gas storage, at 78.6 bcm, was the lowest for the date since 1976 and 6 bcm below the five year average.

Several factors have contributed in recent years to relatively low storage. Some are market-related. There was high demand for gas combined with low supply. Others are weather-related. The summer of 2000 was warmer than usual and the winter of 2001 more severe.

At the beginning of 2001, storage levels were still low and prices averaged about $5/MBtu. By summer 2001, they decreased and stabilised at around $3/MBtu, still fairly high by historical standards. One factor that kept prices high was the need for unusually large refill volumes for underground storage.

Forces favouring storage use – the US and UK experiences

The US market provides an interesting example of the changed role of gas storage in liberalised markets. In the United States, access to storage was mandated by FERC"s Order 636 in November 1993, which foresaw that:

- storage be unbundled from other services and be offered as a distinct service, separately charged;

- customers be offered access to storage capacity or the right to use space in storage reservoirs;

- customers be given the opportunity to sublease any of their contracted storage capacity.

During the 1990s, the operational practices of many underground gas storage sites became much more geared to trading. The following trends have developed:

- There is more storage facilities which can supply at a high rate for a short period, especially in producing regions. Between 1993 and 2000, delivery capacity from high-deliverability storage facilities grew by 62% and the number of sites increased from 21 to 27. The average cycling rate at these sites increased from 1.66 in 1993 to 2.1 in 2000;

- Competitive pressures have shut down some small depleted gas fields that were considered to be uneconomical to operate in the new marketplace or which were thought to be unsafe.

- There is more emphasis on inventory management, with more frequent injection and withdrawal through the year. There is also a trend towards "just-in-time" management of storage, with less inventory at the beginning of the season. Many storage owners are minimising inventories in an effort to synchronise their buying and selling activities with market needs[43];

- Interest in market-based rates has increased as competition has developed in the storage field, and companies are asking for market-based tariffs for such storage services;

- A secondary market for storage capacity (or storage capacity release) has emerged.

The UK experience in storage is less advanced than that of the United States for two major reasons. There are only four underground storage facilities in the United Kingdom, with a working capacity of 3.3 bcm, or only 3% of total gas consumption. Moreover, another source of flexibility was readily available in the past, in the form of short-haul pure gas fields in the southern UK Continental Shelf.

Increasing volatility in gas prices has now sharply increased demand for storage services. Competition has come to storage markets. New facilities are being built by various entrepreneurs. No longer is the business the exclusive preserve of Dynegy, the US company that bought out UK BG Storage and of Transco, the transmission company which also owns the LNG peak shaving facilities.

Virtual storage service providers are also emerging. With the introduction of NETA, the new arrangement for electricity trading which replaced the electricity pool in March 2001, and the gradual convergence between gas and electricity markets, gas storage can now be used to arbitrage daily electricity swing.

Storage and liberalisation in Continental Europe

Continental European companies in the 1970s and 1980s built storage facilities to meet seasonal variations in demand under long-term long-haul supply contracts with a high minimum pay. They also sought to bolster security of supply. These two functions remain in the new liberalised gas markets. In

---

43 EIA (2001a).

addition, however, storage is starting to play new roles, such as making use of price arbitrage and serving as a supporting tool to gas trade.

Companies are beginning to use storage in a very different way:

- Inventories are increasingly cycled more than once a year.

- In choosing sites for new storage capacity, preference is given to more flexible facilities such as salt caverns. Although the working capacity of salt caverns is less than in depleted fields or aquifers, the caverns offer high deliverability rates and can be cycled more than once a year.

- "Virtual storage" to complement physical storage is already offered by companies. Virtual storage are backed up by storage facilities or other flexibility instruments belonging to the company.

The EU Gas Directive does not require member countries to provide open access to storage. Access must be granted to storage only when, and to the extent that, it is technically necessary to move gas efficiently into transmission or distribution networks. However, in the European Commission proposal for a new directive[44], access to storage and ancillary services[45] is mandatory. Member states would have the choice between regulated or negotiated access to storage and ancillary services.

EU countries have in fact chosen different approaches and are at different stages in providing non-discriminatory access to flexibility instruments. In the UK, for example, storage capacity is sold in regular auctions while "virtual" storage can be bought on the spot market. In Italy and Spain, access to storage is regulated. Most other EU countries offering access to storage have opted for negotiated access.

Storage and other flexibility services or instruments offer a competitive advantage to a gas supplier. They lower costs, facilitate balancing and allow provision of greater flexibility and security of supply for customers. Storage is one of the several means to achieve flexibility and will eventually compete with other flexibility services, such as supply swing, interruptibles, spot and futures markets. To the extent that competition develops between providers of flexibility services, there should be little need to regulate access to them.

---

44 CEC (2001c).
45 Ancillary services means all services necessary for the operation of transmission and/or distribution networks and/or LNG facilities, including storage facilities and equivalent flexibility instruments, load balancing and blending.

ELECTRICITY-GAS ARBITRAGE[46] AT HUBS

Before the liberalisation of the power sector, demand for gas in power generation was mainly a matter of price optimisation of different input fuels on a long-term basis. Utilities had a portfolio of power plants which each had its place in the "merit order" for dispatching power. This ranking was a function of operating costs, efficiencies and fuel prices. In the order of merit, nuclear, hydro and lignite were generally used as base-load energy sources and coal was generally used for base- or middle-load. Depending on relative pricing, gas was used for middle- and peak-load. The high efficiency and low capital costs of gas-fired CCGT plants mean that they can also compete for base-load if gas prices allow.

With the increased use of gas for power generation and the opening of the electricity grid to competition, gas demand for power generation becomes more price elastic in the short term as the competitive electricity market offers short-term incentives to take gas or not, according to its price. Power producers can resell gas they have purchased under long-term contracts (if contracts permit). They can sell the gas at the current market price and buy electricity from the grid, if that gives a higher yield than using the gas to produce power.

In a competitive gas and electricity market, the operator of a gas-fired power plant can optimise his operations according to what is known as the "spark spread". The spark spread is defined as the difference, at a particular location and at a particular point in time, between the fuel cost of generating a MWh of electricity and the price of electricity. It is calculated as the difference between the product of the gas price and the heat rate of a power plant (a measure of thermal efficiency) used to generate the electricity less the spot price of electricity at that location. As a result, a positive spark spread indicates the power generator should buy electricity rather than make it.

Arbitrage between the electricity and gas markets functions this way: when the market price of electricity is higher than the price of gas at the power plant, plus variable power production costs and taking into account the thermal efficiency, the power generator will generate electricity from gas. In the opposite case, he will produce from another energy source or buy the electricity on the

---

46 Arbitrage covers three main situations:
 – Time arbitrage is trading the difference of gas prices at different times via spot and futures prices.
 – Geographical arbitrage is trading the difference in the price of gas on different markets. Examples of geographical arbitrage include trading the price difference between natural gas in the UK and Belgium (Zeebrugge) or transatlantic arbitrage of LNG between LNG exported to Europe or to the United States. This calculation will include the cost of transportation of LNG to the United States.
 – Form arbitrage is exploiting the difference in the value of gas in different markets, electricity and spot gas markets. Gas-electricity arbitrage is the most common example of form arbitrage.

spot market[47]. He may interrupt his own production and sell gas instead of burning it. The market price of spot gas is, therefore, increasingly determined by the spot price of electricity.

In the United States, suppliers of natural gas to electricity generators increasingly track the price of power at different locations in real time. When the price of electricity rises at one location, they try to sell more gas into a market nearby, or they transport gas to a particular generator and make an arrangement with the generator to produce more power. In this case, the gas supplier may arrange to sell the power himself – a practice known as "tolling". It occurred in the United Kingdom during the second dash for gas, after 1995, when producers saw that they could earn more from their gas by converting it into electricity than by selling it on the spot market. They arranged with independent power producers to "toll" their gas and take the electricity receipts in exchange for a tolling charge.

In the United States, the complexity of the deregulated gas market and its growing interrelationship with electricity markets have increased the need for coordination among market participants. In addition to dealing with production problems, timely additions of natural gas pipeline capacity and other infrastructure will require coordination among pipeline companies, consumers, the FERC and state regulatory bodies[48].

Box 6: Gas-Power Arbitrage and Prices in the United Kingdom [49]

In the United Kingdom, wholesale power prices have decreased by 20% to 30% since the introduction of the New Electricity Trading Arrangements on 27 March 2001. The decrease reflects more competition in the electricity market, as well as changes in input fuel costs, the effect on new entrants and the erosion of market concentration.

NETA replaced the Electricity Pool of England and Wales[50], in which half-hourly prices had been set by generators, with no scope for bidding by

---

47 In this case, he also needs to take into account the cost of not using his gas plant.
48 EIA (2001b)
49 Source : World Gas Intelligence, April 3, 2002 and 28 November 2001, European Gas Markets, 30 March 2001, The Utilities Journal, March 2002, Energy Trends, DTI, United Kingdom
50 The electricity spot market in England and Wales – the Pool – was a compulsory trading mechanism for generators and suppliers, regulated by its members and operated by NGC. It was mandatory. Generators were obliged to sell their production to the Pool and electricity buyers to buy from it. The Pool set prices for energy for each half-hour period on the basis of a daily day-ahead auction. Generators submitted bids specifying the capacity available for the next day and the price at which they were willing to sell output from each capacity unit. Bids were fixed for the day; in other words, the same prices applied to all half-hour periods. With limited exceptions, there was no demand-side bidding. Bid prices contained several terms such as a fixed start-up rate, a no-load rate for each hour that the unit was running at its technical minimum and various energy rates for different loads. The Pool combined the bids to construct an unconstrained merit order of generating plants that minimised the cost of serving the scheduled demand for each period.

any other players. Under NETA, over 150 companies, including large end-users and small generators, may participate in the marketplace.

NETA has had a radical effect on the gas side. Before its inception, gas and power markets had hardly been intertwined, except that gas has been increasingly used for power generation, going from zero in 1990 to 30% in 2000. With NETA, generators have an incentive to arbitrage between gas and power markets.

The first signs of gas-power arbitrage emerged in 2001, when generators sold gas back into the gas market to seize price differentials. Many gas supply contracts have been renegotiated by power generators to allow delivery at the National Balancing Point, instead of at the beach or the plant gate, and to allow full re-sale rights.

With gas prices at a fairly high 20 to 30 p/therm in 2001, new combined-cycle gas turbines are vulnerable to a margin squeeze, since their fuel costs are not covered by the electricity market price.

Higher gas prices in 2001 have brought a halt to the rising trend in gas use at power stations despite the fact that new gas-fired stations have come on stream in 2000 and 2001. In 2001, supply from gas-fired power plants fell by 2%, while supply from the coal-fired power stations of major power producers rose by 8.5%.

As a consequence of the wholesale power-price decline in 2001, several plants were mothballed before the new UK financial year (April 2002), in order to avoid grid-connection fees and other annual charges. The US company TXU mothballed two 189 MW coal-fired units at High Marnharm and a 333 MW coal-fired unit at Drakelow, representing over 20% of its 2,914 MW of UK capacity. In February 2002, the electricity company AES mothballed its 363 MW coal-fired plant at Fifoots Point. International Power mothballed half the capacity at its 500 MW gas-fired power plant at Teeside.

Although these are all market-related incidents, concerns are emerging in the UK about their longer-run effects on reserve capacity and security of supply, especially in peak periods. Today it is not a problem because the UK has 20%-30% over-capacity in electricity generation, equal to 15-22.5 GW. This overhang looks set to shrink over the next 15 years, since old plants with about 14 GW of capacity are due to be decommissioned.

Given the existence of high generation over-capacity and growing competition in liberalising power markets, the situation in Continental Europe is converging with that of the UK. In 2001, Belgian gas prices were less competitive than usual, and electricity utilities in Belgium reduced their gas off-takes by 5.8% to the benefit of coal and oil.

Since most of the growth potential for natural gas lies in power generation, gas prices to generators will need to be competitive with the electricity price if this demand potential is to be realised. Europe's over-capacity in power generation and the resulting depressed price of electricity in some European countries means that for some time it will not be economic to build new gas-fired capacity.

5

THE ROLE OF LNG
IN FLEXIBILITY

INTRODUCTION

In the past, the liquefied natural gas (LNG) market was characterised by rigidity: long-term Take-or-Pay contracts; liquefaction capacities booked under long-term contracts; almost no supply flexibility and no ships available for released spot volumes.

Market deregulation in many East Asian and European countries is changing the LNG market, bringing new challenges and opportunities to both buyers and sellers. The business is moving away from long-term bilateral contracts towards a system that is more flexible and responsive to market signals. Major new trends include the growing short-term market with more spot transactions and the increasing, albeit still embryonic, globalisation of the LNG business.

INTERNATIONAL LNG TRADE

Global trends

The LNG market is still small compared with the oil market. Total LNG trade was 146 bcm in 2001 (141 bcm in 2000) corresponding to less than 3 million bbl/day, with twelve importing and twelve exporting countries. LNG now represents 22% of the world's total cross-border gas trade.

In 2001, global LNG trade increased only 3%, much less than in former years. A weak global economy, the aftermath of 11 September and mild winter weather limited world LNG growth. However, trade has doubled in the past decade and is expected to experience strong growth in the coming years, with LNG trade expected to rise to approximately 220 to 270 bcm a year by 2010[51].

LNG production requires huge infrastructure investments and long lead times. Up to now, the LNG trade has been built on contracts of 20 to 25 years duration. Take-or-Pay (ToP) and Ship-or-Pay (SoP) arrangements frequently

---

51 Valais M., Chabrelie M.F., and Lefeuvre T. (2001).

Figure 9: Evolution of LNG Trade

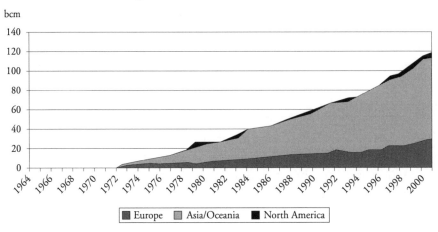

bcm

Source: CEDIGAZ.

cover 100% of the contracted quantity. This was a consequence of the industry's capital-intensive structure. Investments were high, on the order of $5 billion for a two-train 6Mt/year LNG project for liquefaction, ships and regasification. The cost of liquefaction and LNG transportation made up a high proportion of the market value of the delivered gas. Because of the tight economics, there was practically no possibility for the parties in an LNG chain to do business with others. In order to secure capital, LNG purchasers agreed to sign rigid contracts, with ToP obligations covering close to 100% of the contractual quantity and very long-term commitments, typically twenty to twenty-five years. But recent cost reductions in the gas chain and especially in the transportation and liquefaction segments have made it possible for LNG sellers not to bind all their production to fixed counter-parties under long-term contracts and to sell spot to other partners.

Technological progress achieved in the past decades has led to a sharp decrease in investment and operating costs all along the LNG chain. Among other things, this has made LNG more competitive with piped gas. The average unit investment for a liquefaction plant dropped from some $550 per tonne per year of capacity in the 1960s, to approximately $350 in the 1970s and 1980s, $250 in the late 1990s. For projects begun today, the price is slightly under $200 (all in current dollars)[52].

---

52 Ibid.

Regional trends

So far LNG markets are regional. In 2001, Asian importing countries took 73%, Europe 22% and North America 5% of world trade. Among producing regions, 52% came from the Asia/Pacific region, 23% from Africa, 22% from the Middle East and 3% the Americas.

Table 6: Main Characteristics of Regional LNG Markets in 2001

Region	Country	LNG imports (bcm)	Share of LNG in total gas imports (%)	Share of LNG in gas consumption (%)	LNG buyers' position/ Power
Asia	Japan	78	100	97	Dominant position
	Korea	21	100	100	Monopoly
	Taiwan	7	100	100	Monopoly
North America	USA	6.7	6	1	Full competition
Europe		33.3	11	7	Emerging competition

Source: Cedigaz, BP, IEA.

■ Asia

The Asian market, which imported 106 bcm in 2001, represents 73% of world LNG trade, with Japanese gas and electricity companies buying three-quarters of the regional total. Asia has played a major role in the increase of world LNG demand. Between 1985 and 2001, about 70% of the increase in world LNG demand came in this region. LNG plays a major role in Asia in diversifying sources of energy and reducing air pollution when it replaces oil and coal.

The Asian LNG markets were developed on the basis of dedicated long-term supply. Historically, LNG was priced higher in Asia than in Europe and North America where the competitive conditions were different. In Europe and North America, LNG has to compete with pipeline gas. In Japan, it competed against light crude oil in base-load power generation.

Japan was the first Asian country to import LNG, and is currently the biggest importer in the world, taking 78 bcm in 2001 from eight supplying countries. South Korea started to import LNG in 1986 and is now the second largest importer in the world, taking 21 bcm in 2001. Taiwan is a more recent Asian importer, having started in 1990. Next to come are China with the Guangdong project and India with several ongoing LNG receiving terminals projects, two of them under construction.

LNG in Asia is at a transition point, with significant changes ahead in Japan, South Korea and Taiwan, as well as in the new emerging LNG markets. The market was shaken by the Asian financial crisis in 1997-1998, when Japanese and Korean buyers found themselves stuck with volumes contracted well above their requirements. This brought home the need to reduce the length of contracts and increase their flexibility. In this respect, changes are becoming visible with the current renegotiation of contracts (see p. 106). Growth has started again; Asian LNG demand is expected to reach 107 to 128 Mt (144 to 173 bcm) in 2010.

■ Europe

Europe imported 33.3 bcm of LNG in 2001. That amount represents only 7% of European gas consumption. But the share is much higher for some Mediterranean importers. In Spain, for example, LNG accounts for 54% of gas supplies.

In Europe, LNG competes with pipeline gas, and both compete with other fuels. European buyers choose LNG either to diversify their gas portfolio or to supply areas far from the main gas grid. The major suppliers have been Mediterranean countries: Algeria and, to a lesser extent, Libya. In 1999 and 2000, additional LNG supply started from further afield: Nigeria and Trinidad and Tobago.

Continuing increases in demand and the liberalisation of European gas and power markets are leading to new opportunities for LNG, especially in the Mediterranean. Despite strong competition from pipeline suppliers, LNG deliveries to Europe are expected to rise steeply, reaching 39-48 Mt/year (53 to 65 bcm) in 2010, approximately twice as much as in 2001. New LNG receiving terminals are planned in France, Italy, Portugal, Spain, Turkey and the United Kingdom, which may eventually resume importing LNG. Competition among suppliers is likely to facilitate the development of new LNG projects in Europe. The number of short-term LNG purchasing contracts is growing, as European buyers benefit from surpluses available in the Middle East and Africa.

■ United States

In the past few years, US LNG imports have increased again, after stagnating at a very low level for a long time. Indeed, two terminals on the East Coast have been mothballed since 1979. Gas prices rose sharply in 2000 as demand increased and supply was very tight. As LNG supply costs have decreased, and US gas prices have been high, LNG has become competitive in the US again. LNG imports rose to 6.4 bcm in 2000, 40% higher than in 1999 and 6.7 bcm in 2001, or 1% of total US gas supply. The slower increase in 2001 was due to safety concerns after 11 September and to lower gas prices in the US. The US received only 12 cargoes during the fourth quarter of 2001, compared with 75 during the first three quarters.

The two LNG receiving terminals in operation at Lake Charles, Louisiana, and Everest, Massachusetts, are increasing their throughput capacity. The two terminals that have been mothballed for many years, at Cove Point, Maryland, and Elba Island, Georgia, are being brought back into operation, and will even be enlarged.

LNG deliveries from Algeria and from Trinidad and Tobago are received under long-term contracts. Spot cargoes have been imported from Qatar, Nigeria, Australia, Oman, Indonesia and Abu Dhabi. The spot sales market was very active in 2000 and in 2001, but since mid-2001, price trends have been unfavourable. The current breakeven for US LNG imports is in the range of $3 to $3.5 per MBtu. Over the long term, LNG is likely to be an attractive option for increasing gas supplies to the US. Imports could reach 15 to 19 Mt/year (20 to 26 bcm) by 2010.

DEREGULATION OF ASIAN AND EUROPEAN LNG MARKETS

Downstream gas markets in Japan and Korea are in a state of change, as governments introduce competition (see Chapter 2). Current deregulation trends in the electricity and gas industries are producing a different business environment from the past, and many uncertainties lie ahead. Buyers in Japan and Korea are now more reluctant to enter into rigid long-term contracts. In both countries, third-party access to LNG facilities is due in 2003.

In this new environment, Japanese LNG purchasers are asking for more flexible contract terms based on each purchaser's individual requirements. According to the Japanese company Chubu Electric[53], LNG buyers now seek a portfolio of contracts, ranging from long- to medium-term to short-term, as well as spot transactions, which will allow them to cope with their demand patterns

---

53 Kuroyanagi (2001).

and with uncertainties ahead. When renewing existing contracts, LNG purchasers are seeking periods of 5 to 15 years rather than 20 to 25, more flexible volumes off-take and purchases on a fob basis[54]. The newly-signed contracts between Japan and Malaysia (see p. 106) are the first sign of a change in LNG marketing in Asia.

In Europe, the EU Gas Directive provides for third-party access to LNG terminals, either regulated or negotiated. TPA frameworks are now in place in European countries. Italy recently adopted negotiated TPA to LNG terminals. The Italian regime applies only to newly-built terminals; access to the transmission grid and to the existing terminal is regulated. In France, the Gas Directive has not yet been transposed into national law, but Gaz de France took the initiative to publish access conditions and charges in January 2001. Belgium initially opted for negotiated access but is now moving to regulated TPA. Belgian tariffs and access conditions have not yet been published. In Spain, a Royal Decree adopted in August 2001 provides for regulated TPA to LNG terminals. Revised tariffs were published in February 2002. As emerging markets, Greece and Portugal obtained exemptions from implementing the EU Gas Directive provisions up to 2006.

These legal provisions are all quite recent, and they have yet to effect any major changes in access to European LNG terminals. Nevertheless, third-party access is increasing. In France, Distrigas gained access to the Montoir-de-Bretagne terminal; TotalFinaElf to Zeebrugge in Belgium, Edison Gas to La Spezia in Italy, while BP Global LNG, Cepsa, Shell and ENI have all made deliveries to the Spanish LNG terminals.

LNG is playing a major role in the liberalisation process as many new entrants have chosen to build their own LNG terminals and to import gas directly. New players, principally electricity companies, are entering the LNG business. They include Union Fenosa, Iberdrola, Cepsa and BP in Spain, Edison and the BG Group in Italy and TFE in France and Spain.

As competition in their markets increases, European LNG buyers have begun seeking contracts that are more flexible than the traditional 20-25 year ToP contracts pegged to oil prices. LNG spot sales have rapidly increased since 1997. Spot sales amounted to 3 bcm in 2001, most of them for the Spanish market. Short-term contracts are also being signed with suppliers (see p. 106).

---

54 Free-on-board (fob): Under a fob contract, the seller provides the LNG at the exporting terminal and the buyer takes responsibility for shipping and freight insurance.

New arrangements between gas suppliers and buyers are emerging. Algeria's Sonatrach and Gaz de France have established a 50/50 joint venture, Med LNG & Gas, to carry out short- and long-term LNG sales principally to the European and North American markets. Med LNG & Gas will be free to sell up to 1 bcm of gas per year on the market of its choice, provided the price charged is higher than the price quoted in the other Gaz de France/Sonatrach transactions.

NEW TRADING ARRANGEMENTS

LNG trading is increasingly showing more flexibility in contract terms and destination, as well as in spot and short-term transactions, LNG swaps and arbitrage transactions.

Swap agreements

Swap agreements - which are not new in the gas industry - are developing. A swap is usually an exchange of volumes not driven by price considerations. Two quite different types of swaps can be observed. The first is aimed at shortening delivery routes; here there are sellers at two different locations and buyers at two different locations. The second is intended to bridge differences in demand pattern between purchasers. It generally occurs when there is an over-commitment to take or pay or contractual entitlements not used in one place, and a lack of supply in the other with a mirrored situation later on. The first type cuts transportation costs and frees up shipping capacity; the second type leads to increasing market flexibility by smoothing out differences in the timing of demand between two purchasers.

Examples of the first type of swap agreements include those among Spain, Algeria and Trinidad. In 2000, Spain's Gas Natural became the first European LNG buyer to resell LNG to the US market. That gas has been sold to Gas Natural by Atlantic LNG of Trinidad. At the same time, Algerian LNG dedicated to the US was delivered to Spain, reducing shipping charges for all parties. These swaps developed in 2001 into a more permanent arrangement with the signature of a contract among Sonatrach of Algeria, Gas Natural of Spain, Tractebel LNG North America in the US and Belgium's Distrigas.

Companies with interests on both sides of the Atlantic have an advantage over others as they can react fast to any market opportunity. This is currently the case of Gas Natural and Tractebel LNG, and it may soon be the case for BP,

BG and Shell, all of which are positioning themselves on both sides of the Atlantic.

An example of the second type of swap agreement took place in 2000 and 2001. In 2000, Japan and Korea swapped cargoes with Taiwan whose peak requirements do not coincide with theirs. Taiwan gave the volumes back in 2001.

New contractual terms

There have also been major changes in the terms of contracts. In the past, most contracts were for a 20-25 year period and were almost entirely subject to ToP or SoP obligations. New long-term contracts have a shorter duration of 15 years. Also medium-term contracts, of 5 to 8 years, are becoming more common. ToP obligations have been relaxed.

In Europe, several shorter-term contracts have recently been signed:

■ Oman LNG signed a contract with Shell for the sale of 700,000 tons of LNG per year from February 2002 over five years. The LNG is intended for Spain.

■ BP signed a three-year contract to buy up to 750,000 tons of LNG per year from Abu Dhabi's Adgas, starting in 2002. The volume is very flexible; it could be 300,000, 500,000 or 750,000 tons per year. The sales are on a fob basis.

■ In May 2001, Qatargas signed a contract with Gas Natural for the supply of 12-13 fob cargoes of LNG per year from October 2001 to July 2009 and for the supply of 12-13 ex-ship cargoes[55] per year from July 2002 to July 2007, with a possible extension to 2012.

In Asia, two recent transactions represent significant deviations from traditional LNG marketing:

■ In February 2002, the first new contract with increased flexibility was signed between three Japanese gas utilities and Malaysia. Tokyo Gas, Osaka Gas and Toho Gas, undertook to buy from Malaysia LNG (Tiga) a combined 680,000 tons of LNG per year for 20 years, and an additional 340,000 tons for the single year beginning April 2004. The single-year component of the contract is to be updated annually, with volumes specified a year in advance. This combined short-term/long-term contract effectively

---

55 Ex-ship: Under an ex-ship contract, the seller has to deliver LNG to the buyer at an agreed importing terminal. The seller remains responsible for the LNG until it is delivered.

provides 40% volume flexibility, instead of the 5%-10% available under conventional contracts.

■ Also in February 2002, the Japanese companies Tepco and Tokyo Gas agreed in principle to renew their long-term contract with Malaysia LNG (Satu) for 15 years, starting in 2003. The current 20-year contract for 7.4 million tons of LNG per annum expires in March 2003. In the process of preparing the new contract, Tepco and Tokyo Gas gained key concessions on flexibility. Instead of all cargoes being sold on a delivered basis, after 2003 up to one quarter can be lifted fob Malaysia. There is also a buyer option to cancel up to two-thirds of the 1.8 Mt/year fob volume, or 1.2 Mt, every four years, with one year's advance notice. Press reports have said that the buyers have secured a price cut of approximately 5%.

Besides altering the duration and volume of the contract, a buyer can also achieve greater flexibility in its LNG supplies including the length of the build-up period, make-up and carry-forward rights.

Arbitrage: The development of Transatlantic LNG trade

A Transatlantic LNG market is starting to emerge, with spot trading, physical arbitrage between European and US markets and an optimisation of shipping costs.

In the US market, LNG cargoes can always be sold at short notice, within spare LNG receiving terminal capacity. High prices in the United States during the winter of 2000-2001 made the US market very attractive to sellers. Simultaneously, Middle East producers had idle supply capacity in the wake of the Asian financial crisis. The Middle East is well located to ship LNG to all three regional markets. In December 2000, prices at the Henry Hub reached $10/MBtu, and remained in a range of $4 to $9 throughout the first half of 2001. After negotiations with their suppliers, some European term buyers of LNG redirected cargoes to capture the economic benefit of high US prices.

The 2001 US gas-price boom affected European prices. LNG supply redirected from Europe was replaced by cheaper spot purchases of UK gas from the Interconnector. The short-term effect of this development, combined with higher Continental gas prices, was an increase in UK prices, which reached 32 p/therm ($4.75/MBtu) at the National Balancing Point in January 2001.

The within-day price went even higher, to around 45-to-50 p/therm ($6.70 to 7.40/MBtu).

But a fall in US prices in spring and summer of 2001 narrowed the scope for LNG arbitraging, and even reversed the flow. LNG cargoes contracted for the US were redirected to Europe. This trend was reinforced after 11 September, partly because of a ban on LNG imports to Boston for safety reasons, and because of the economic recession that followed the terrorist attacks.

Spot sales

The global LNG spot market has been booming since 1999. Spot sales rose to 11.41 bcm in 2001, an increase of 51% over 2000. In 2001, short-term LNG trading represented 8% of all LNG trade.

The evolution of the spot LNG market[56] is indicated in the following table:

Table 7: LNG Spot and Swap Transactions - 1992 to 2001
By Exporting Country - bcm

Exporters	1992	1993	1994	1995	1996	1997	1998	1999	2000	2001
Abu Dhabi	-	-	-	1.43	1.39	0.08	0.34	0.65	0.64	0.31
Algeria	0.53	0.49	0.59	0.35	-	0.60	0.45	1.33	1.38	2.64
Australia	-	0.34	0.58	0.67	0.27	0.30	0.38	0.30	0.45	0.21
Brunei	-	-	0.30	0.08	-	-	-	-	-	-
Indonesia	0.23	0.24	0.38	0.53	0.60	0.28	-	0.38	1.18	1.91
Libya	-	-	0.05	-	-	-	-	-	-	-
Malaysia	0.30	0.53	0.45	0.23	0.08	-	-	0.08	0.08	0.52
Nigeria	-	-	-	-	-	-	-	-	0.37	1.22
Oman	-	-	-	-	-	-	-	-	0.60	0.58
Qatar	-	-	-	-	-	0.39	0.95	1.60	1.98	2.62
Trinidad	-	-	-	-	-	-	-	0.39	0.92	1.40
Total	**1.05**	**1.59**	**2.34**	**3.27**	**2.33**	**1.64**	**2.12**	**4.72**	**7.58**	**11.41**

---

56 Spot transactions and short-term contracts of less than one year.

By Importing Country - bcm

Importers	1992	1993	1994	1995	1996	1997	1998	1999	2000	2001
Belgium	-	0.23	0.08	0.15	-	-	-	-	-	0.07
France	-	-	-	0.87	0.23	-	-	0.08	0.08	0.43
Italy	0.53	0.26	0.20		-	-	0.12	0.54	0.48	0.38
Japan	0.38	0.39	0.08	0.08	0.15	0.28	-	0.15	0.32	2.22
Korea	0.15	0.45	1.05	0.90	0.68	-	0.08	0.31	1.47	1.85
Portugal	-	-	-	-	-	-	-	-	0.08	-
Spain	-	0.27	0.94	1.05	0.98	0.99	0.83	1.69	1.43	2.20
Taiwan	-	-	-	-	-	-	-	-	-	0.08
Turkey	-	-	-	0.23	0.08	-	0.58	0.30	-	-
US	-	-	-	-	0.23	0.30	0.53	1.66	3.73	4.18
Total	**1.05**	**1.59**	**2.34**	**3.27**	**2.33**	**1.64**	**2.12**	**4.72**	**7.58**	**11.41**

Share of spot in global LNG Trade - %

Year	1992	1993	1994	1995	1996	1997	1998	1999	2000	2001
	1.3	**1.9**	**2.7**	**3.5**	**2.3**	**1.5**	**1.9**	**3.9**	**5.4**	**7.8**

The figures include both spot LNG sales and swap transactions.
Source: PetroStrategies (1992-2000 data), US Department of Energy, International Group of Liquefied Natural Importers[57] and IEA estimates for 2001.

More and more players are selling and buying spot LNG. This phenomenon was triggered by the 1997-98 Asian financial crisis, which generated supply surpluses in the Middle East. This was followed by spot sales to meet peak winter demand in Korea and Spain. The next driving force was the US market with the very high prices of 2000-2001, which led to spot cargoes' being redirected from Europe to US, as well as direct LNG spot purchases. About 4 bcm/year was imported to the US under spot or short-term contracts in 2000 and 2001.

The increase in global spot sales in 2001 resulted partly from exceptional circumstances. The temporary shutdown of the Arun liquefaction plant in

---

57 GIIGNL (2002).

Indonesia gave a real impetus to spot trading. ExxonMobil decided to shut down three of the four fields supplying the Arun plant, after repeated attacks on its workers from separatist rebels. This forced Pertamina, the state Indonesian oil and gas company, to declare a state of *force majeure* at the plant. Indonesia made up most of the shortfall with spare capacity at Bontang; spot cargoes from Bontang to Japan and South Korea amounted to 1.9 bcm in 2001. Arun's Japanese and South Korean customers, Tohoku Electric and Tepco in Japan and Kogas in Korea, acquired the balance from other sources, mainly Qatar, Malaysia and Australia.

Spot volumes in 2001 were produced mainly by Qatar with 23%; Algeria with 23%; Indonesia, 17% and Nigeria, 11%. The spot trade now relies on excess production capacity in newly-built plants in the Middle East. For Middle East producers, as contracts are in the build-up period[58], spot sales are an attractive way to make best use of their plant.

For LNG sellers, short-term trading provides an opportunity to market cargoes that have not been contracted on a long-term basis. This reduces the risk of going into a project without having secured long-term sales for all the planned production. In addition, a trading system may enable sellers to dispose of cargoes that buyers cannot use because of a downturn in demand in their market. From a producer's perspective, spot selling is advantageous, so long as spot LNG prices do not undermine long-term LNG prices and revenues.

PRICING DEVELOPMENT

A more flexible approach to pricing is emerging in the LNG industry, with the adoption of new indices such as electricity pool prices, and the adoption of the so-called "S" curves[59] for LNG prices.

The "S" curve formula was first adopted in 1989, in contracts between Australian LNG suppliers and their Japanese customers. It was adopted last year in a number of European contracts.

---

58 During the term of a contract, there are different stages. The build-up period will typically consist of several steps by which the contractual quantity is gradually increased up to the plateau or peak level.

59 The S-curve is a formula in which the variable portion of the LNG price is adjusted in accordance with the price evolution of a basket of crude oils. Its evolution is linear within a price range (roughly between $21 to $28 per barrel). It shifts upward when the price of oil falls below the floor price. It shifts down when the oil price exceeds the ceiling price. This curve is intended to protect the interests of both contractual parties against highly volatile oil prices.

Price renegotiations in 2001 also included new indices in the price formula, such as electricity pool prices in the formula negotiated between Trinidad and Tobago and Spain's Gas Natural. As gas trading develops and new gas market indices appear, future LNG contracts could be pegged to them.

Middle East producers have developed master agreements for spot LNG trade or transactions concerning just a few cargoes. These standardised framework agreements facilitate spot transactions. They allow cargoes to be re-routed, provided that the LNG producer gets a piece of the benefit. Oman LNG, Adgas in Abu Dhabi, and Rasgas and Qatargas in Qatar are among the suppliers which have such master agreements. Nigeria LNG has signed such agreements with US CMS, Coral, a Shell affiliate in the US, TotalFinaElf and BP.

■ **Regional pricing**

Currently there are three regional LNG markets, each with its own pricing.

In Japan, cif[60] LNG prices are based on a basket of crude oils imported into the country and known as the Japanese Crude Cocktail (JCC). In the past, this "cocktail" was a convenient basis for gas pricing because the main competitor of gas was light crude oils, whose prices are reflected in the JCC.

Prices in European LNG contracts are still predominantly linked to the evolution of gasoil and heavy fuel oil prices over a given period, usually six months to one year. In some contracts, however, other indices, such as electricity pool prices have been included to reflect the new competitive situation of gas in power generation. European LNG contracts are also less rigid than Japanese ones, as they include renegotiation clauses and opportunities to reopen price discussions.

In the US market, LNG prices are generally linked to the Henry Hub prices. Ex-ship prices tend to represent 80% to 90% of the futures prices at Henry Hub, adjusted for the location of the LNG terminal.

These pricing mechanisms result in three different regional price patterns, as indicated in figure 10.

With increasing short-term trading and physical arbitrage between the regional markets, an additional pricing structure is beginning to evolve. Spot prices reflect the supply and demand situation at any one time. In times of glut, spot prices will be below long-term contract prices and in times of squeeze, they will

---

60 Cost, insurance and freight (cif): A cif price means that the cost of transportation, insurance and freight to a given destination are all included in the price. The seller is usually responsible for arranging transportation.

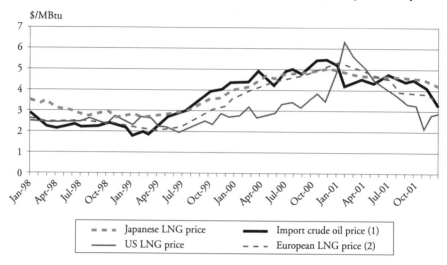

Figure 10: Evolution of LNG Prices in the United States, Europe and Japan

- - - Japanese LNG price   ——— Import crude oil price (1)
——— US LNG price   - - - European LNG price (2)

(1) average import crude oil price in Japan.
(2) Algerian LNG received in France.
Source: US DOE/EIA, World Gas Intelligence, IEA.

be above. As spot transactions develop, differences in timing, volume and duration of LNG contracts will be reflected in their price.

DRIVING FORCES IN THE DEVELOPMENT OF AN LNG SPOT MARKET

Given the increase in LNG spot sales, the question arises: is a global LNG market developing?

A spot market, in the sure sense of the term, aims to match short-term surpluses held by a supplier with unfilled demand of the buyer. It is a permanent dynamic or tension between supply and demand for the commodity for immediate delivery. The price of the goods is continuously quoted, on the basis of supply and demand. A spot market needs the following prerequisites to be fully efficient:

■ Large number of players, without any leading player who could manipulate prices.

■ A fluid market with little or no bottleneck; transport of the commodity should not hinder its availability, in place or time.

■ At least one place where the goods can be delivered to fulfil open deals.

LNG spot transactions do not yet constitute a real LNG spot market, in the sure sense of the term. LNG spot transactions are still marginal at 8% of global LNG trade.

The recent growth in LNG spot sales was made possible through spare capacity in liquefaction plants, the availability of at least a few LNG tankers and spare capacity in receiving terminals.

Spare capacity at liquefaction plants

There is a good deal of spare LNG capacity - capacity that is not contracted, or contracted but not taken. Spare capacity was estimated at 17 bcm in 2000 and 12 bcm in 2001, the decrease being mainly due to the seven-month shut down of the Arun plant in Indonesia. This spare capacity exists for several reasons. In newly built plants, it exists because contracts are still in their development stage. In existing plants, spare capacity was produced by de-bottlenecking and in some newer plants - thanks to cost reductions - not all of the initial capacity has had to be sold in advance to guarantee financing.

Most spare capacity is in the Middle East and in Africa. In the Middle East, three new liquefaction plants have been put on stream since 1996, Qatar's Rasgas, Qatargas and Oman LNG, adding a total of 20 Mt capacity per year. In Africa, Nigeria has added capacity of 5.3 Mt per year. All contracts linked to these liquefaction plants are in their build-up stage:

- Abu Dhabi: 4.3 Mt contracted long-term in 2001, out of 5.5 Mt/year capacity (78%).

- Qatar: 10.16 Mt contracted long-term in 2001, out of 13.4 Mt/year capacity (76%).

- Oman: 3.29 Mt contracted long-term in 2001, out of 6.6 Mt/year capacity (50%).

- Nigeria: 4.4 Mt contracted long-term in 2001, out of 5.3 Mt/year capacity (83%).

By comparison, in 2001, the other exporting countries contracted from 84% to 93% of their total liquefaction capacity on a long-term basis. Australia and Brunei contracted their total capacity.

Current spare capacity is the result of a build-up of long-term contracts and will therefore disappear when the contracts have reached their plateau level. Most new greenfield investments are still linked to long-term contracts, and capacities are seldom deliberately earmarked for spot trade.

On the other hand, debottlenecking of existing plants will continue, while new plants and extensions to existing ones will be built around the world, regularly adding spare capacity not yet booked. Five plants, Atlantic LNG in Trinidad and Tobago, Nigeria LNG, Qatargas, Oman LNG and Australia's North West Shelf, are currently being expanded, and many owners of liquefaction plants are discussing additional expansions.

There is also a new tendency by LNG producers to build plants without having full capacity booked under long-term contracts. This development is due to the substantial fall in building costs, especially for expansions. These savings allow the financing of the project to be secured by long-term contracts without committing the plant's total capacity. For example, both Qatar's RasLaffan LNG and Oman LNG committed to project construction without full-capacity sales. This had two benefits. The principal buyer, in both cases Korea Gas, did not have to wait for the seller to line up other buyers before the projects were begun. Second, the projects made subsequent sales based on not-yet committed capacity - Oman LNG to Japan and to the Indian Dabhol project, and RasGas to India's Petronet project. Malaysian producers are also building an extension to their LNG plant, Tiga, without full commitment for the production.

Shipping issues

The current shortage of LNG carriers is restricting spot LNG trade. This situation has persisted since 1999, even though a number of new LNG carriers have entered service since then. The fleet consisted of 128 vessels at the end of 2001, but only five or six are available on an *ad hoc* basis for spot trade. This is a serious bottleneck. However, it is likely that there will be a surplus of LNG tankers in a few years. At the end of January 2002, shipbuilders had no less than 53 firm orders for LNG tankers on their books, with options outstanding on a further 23. Until these ships are delivered, LNG transport will be tight, but by 2004, the situation will ease up and many ships should be available for spot transactions.

Almost all the LNG vessels in use, whether chartered by sellers or buyers, are dedicated to specific projects and routes on a long-term basis. But a new trend is emerging with so-called "free ships", that are not engaged in any specific business and so can provide the flexibility required for spot trading.

Over the past two years, orders for new LNG tankers have increased as major oil companies, independent ship owners and LNG purchasers responded to the lure of low prices. LNG ship prices have fallen dramatically, down to an average of $170 million for a 135,000 cm vessel, from $250 million ten years ago. Companies like Shell and BP have placed orders for LNG carriers which are not designated for specific import or export contracts. Independent ship owners have also placed orders for LNG carriers which are not chartered for long-term trade.

On the purchasers' side, a number of Japanese, US and European electricity and gas companies have decided to order LNG vessels, for fob contracts and for spot purchases. This is not simply to take advantage of low ship prices, but is in line with a policy to make procurement more flexible and better matched to each market's individual needs. Having the means to arrange delivery allows buyers a much freer choice of suppliers. Several buyers have invested in more shipping than they will need in order to lift their volumes contracted on a long-term basis. They are now in a position to purchase additional cargoes of LNG or to use their ships to deliver LNG cargoes to other markets.

Regasification terminals

In 2000, the major bottleneck to expanded LNG trade in the United States was limited importing capacity. With the scheduled reopening in 2002 of two previously mothballed plants, Cove Point, Maryland, and Elba Island, Georgia, US importing capacity will reach nearly 32 bcm/year at the end of 2002. There are plans to enlarge the capacity at existing plants. There are also many new regasification projects planned or under consideration in North America: in the Gulf of Mexico, North Carolina and Florida. To avoid siting problems for the new terminals, especially in California, there have been proposals to site them just outside US borders, notably in Baja California, Mexico, and in the Bahamas. There is also a project to build an offshore regasification terminal in the Gulf of Mexico, and to convert LNG tankers by installing regasification facilities onboard them.

In Europe, some LNG terminals have no spare capacity (La Spezia in Italy and Huelva in Spain), others have limited spare capacity (Fos-sur-mer in France, Barcelona and Cartagena in Spain), while Zeebrugge in Belgium, Montoir-de-Bretagne in France, and Marmara Ereglisi in Turkey, do have available spare capacity. In all countries with existing LNG terminals except Belgium, there are projects to build new terminals, some of which are more likely to be realised than others: two projects in France at Fos and Le Verdon; eight projects in Italy at Marina di Rovigo, Brindisi, Taranto, Vado Ligure, Muggia-Trieste, Corigliano, Lamezia Terme and Rosignano Marittimo; three projects in Spain at Bilbao, El Ferrol and Valencia; two in the United Kingdom at Milford Haven and on the Isle of Grain. A new terminal in Turkey at Izmir has recently been completed.

The legislation on access to LNG terminals will be the determining factor for the building of new terminals. Current proposals within the frame of the EU Gas Directive require LNG receiving terminals to offer third-party access. The financing of the new terminals remains, however, an open issue. Italy is the first country to have adopted a specific legislation encouraging the financing of new LNG terminals. It allows a priority of access to the sponsor of the new terminal and a higher rate of return than is now applicable to the transmission network (see Annex on Italy).

The current eligibility threshold for choosing one's supplier in EU member states is 25 mcm/year, or 40,000 cm of LNG, about the size of a "small" tanker. A new entrant with a small gas market will have to pay high storage costs if he wants to supply his customers with LNG.

Japan has ample spare capacity at its regasification terminals. South Korea is going to expand the capacity of its two existing terminals and build a new one.

A global LNG market?

There is now some, but not a great deal of, spare capacity in liquefaction plants and regasification terminals. The number of free tankers is growing. More players are participating but the total is still limited at around sixty. Traditional long-term LNG contracts are gradually being complemented by LNG transactions that are more flexible in timing and location. These spot transactions, which represented 8% of global LNG trade in 2001, will develop further but they will not soon replace long-term deals entirely, as these deals will still be needed to secure new-project investment. In any event, chances are that a global LNG market, such as those for oil or coal, will gradually emerge.

IEA COUNTRIES ANNEXES

IEA COUNTRIES ANNEXES

The following annexes detail seasonal gas demand in each IEA country and the way seasonal gas demand is balanced by traditional flexibility tools. The information contained here refers to volume flexibility and is based on monthly information available in an IEA database. It therefore refers to seasonal, and not to hourly or daily, flexibility.

Information on each country is presented in the following form:

1. Demand side

- size of temperature-sensitive sector,
- seasonality of gas demand,
- gas in power generation and multi-fired electricity generation,
- interruptible customers and fuel-switching capabilities.

2. Supply side

- reserves,
- gas supply structure,
- production swing,
- flexibility in imports.

3. Storage

- underground gas storage,
- LNG receiving terminals,
- stock changes/load balancing.

4. Regulatory framework

- security of supply,
- access to transmission and storage,
- development of gas hubs.

The following definitions have been adopted:

- The *temperature-sensitive sector* refers to residential and commercial customers.

- *Seasonality in gas delivery* is the ratio between gas consumption in the peak and the lowest month of a given year.

- *Flexibility in gas supply* (production and imports) is the swing, or the maximum monthly delivery divided by the average monthly delivery in a given year.

- *Reserves* means proven reserves.

- *R/P* stands for the reserves-to-production ratio in years.

The table on fuel-switching capacity in the power sector indicates: the share of thermal power plants in the total electricity capacity of each IEA country; the share of multi-fired power plants; and the plants that can switch from or to natural gas.

Sources of information

The following sources of statistical information have been used for the graphs, unless mentioned otherwise:

- IEA Natural Gas Information 2002, Paris, OECD.

- IEA Electricity Information 2002, Paris, OECD.

- IEA monthly database.

- Natural Gas in the World, CEDIGAZ (2001), Rueil Malmaison.

For consistency reasons, data refer to the year 2000, unless stated otherwise.

Units

- *mcm* stands for million cubic metres,

- *bcm* stands for billion cubic metres,

- *mcfd* stands for million cubic feet per day,

- *Mt* stands for million tons,

- *GW* stands for gigawatts.

AUSTRALIA

- The main consuming sectors are industry and power generation. The residential and commercial sectors represent only 19% of total consumption. Seasonal fluctuations in gas demand are less pronounced than in other OECD countries.

- Australia has developed four underground storage facilities, which help to cover fluctuations in gas demand.

- Australia is rich in gas resources and is a big LNG exporter.

- To date, the Australian gas market has, to a large extent, developed as separate markets in individual states, with few interconnections and exchanges.

GAS DEMAND

- **Share of gas in TPES:** 18% (2000)

Gas consumption reached 22.5 bcm in 2000. Unlike many other Asia-Pacific countries, where the bulk of gas use is in the power sector, it is industry that uses the largest share, 38% in 2000. The power sector accounts for 23%. The core sector (residential and commercial users) represented 19% only of total consumption.

To date, the Australian gas market has developed largely as a set of separate markets in individual states, with few interconnections and exchanges. Australia's natural gas reserves are linked to major markets by over 19,000 km of high-pressure transmission pipelines. Most natural gas markets are supplied by a single pipeline that carries gas from a single production centre.

Figure 11: Australian gas consumption by sector in 2000

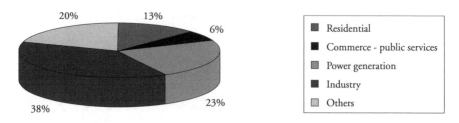

■	Residential
■	Commerce - public services
■	Power generation
■	Industry
▢	Others

■ **Seasonality**

The ratio between gas sales in the peak and the lowest month of the year was 1.6 to 1 in 2000.

■ **Share of gas in the power-generation sector:** 13% (2000)

Table 8: Multi-Fired Electricity Generating Capacity in Australia at 31 December 2000 (GW)

Solid/Liquid	-
Solids/Gas	-
Liquids/Gas	0.16
Liquids/Solids/Gas	0.64
Total Multi-fired	0.80
Total Capacity*	36.51

* From combustible fuels.

2% only of total electricity-generating capacity by fuel is multi-fired. Historically, coal has been by far the fuel of choice in the power sector, with 80% of electricity generation. However, gas is penetrating the sector and gas-fired power plants are expected to account for 17% of electricity generation in 2005 and 20% in 2010.

■ **Interruptibles and fuel switching**

There are very few multi-fired electricity plants in Australia. All gas-fired electricity plants are single fuel-fired.

GAS SUPPLY

■ **Reserves:** 3,530 bcm   **R/P:** 108 years

Australia has abundant gas resources. Unfortunately, these resources are not evenly distributed, the bulk of them are located in offshore Western Australia, far from the major consuming centres.

■ **Gas supply structure:** Indigenous production 100%

Gas production reached 32.7 bcm in 2000.

■ **Supply swing**

Production offers little swing: 119% in 2000.

Figure 12: Australian monthly gas production

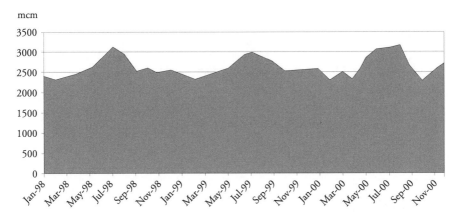

Australia is a net exporter of gas. It exports 35% of its production in the form of LNG through the North West Shelf project located in the Carnarvon Basin. In 1999/2000, LNG production and exports rose to a peak of 7.9 Mt/year, as the contracted sales to Japan were complemented by spot sales to Spain, South Korea, Turkey and the United States. The LNG project currently operates at maximum capacity. Two additional LNG trains are planned at the NWS project, to go into service in 2004. Five other potential LNG projects are under consideration.

STORAGE

Australia has developed four underground gas storage facilities in depleted gas fields and an LNG peak-shaving facility near Melbourne.

Table 9: Undergound gas storage in Australia

Name	Type	Operator/ Number	Working capacity (mcm)	Peak output (mcm/day)
Mondara Field, Perth Basin	Depleted gas field		127	5
Moomba, Cooper Basin	Depleted gas field		623	4
Newstead, Surat Basin	Depleted gas field		234	
Iona Field, Otway Basin	Depleted gas field		260	5.2
Dondenong	LNG		18	6
Total		**5**	**1,262**	**20.2**

- **LNG exporting terminals**

Australia has one large LNG terminal at Burrup, with storage tanks of 260,000 cm of LNG.

Table 10: LNG exporting terminals in Australia

Terminals	Storage tanks (1,000 cm of LNG)	Nominal capacity (mcm/day)	Start-up date
Burrup (North West Shelf)	260	10.1	1989

- **Stock changes**

Figure 13: Australian load balancing

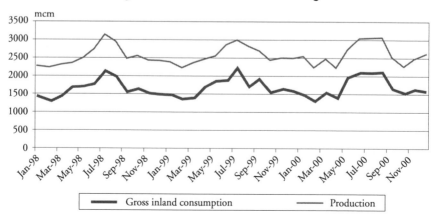

Production offers little swing. Seasonal variations in gas demand are limited and covered by swing in production and storage.

REGULATORY FRAMEWORK

- **Security of supply**

Steps have been taken towards the gradual establishment of a national integrated gas market, which will reinforce flexibility and security of supply by ending the present dependence of each gas market on a single long pipeline.

An explosion at the Longford gas plant in Victoria in September 1998 demonstrated the vulnerability of gas supply. It disrupted gas supplies to Victorian consumers for nearly two weeks. The incident provided a new impetus to gas industry reform.

Since the Longford accident, a number of State Governments have been preparing an agreement to share gas supplies across borders[61] in case of emergency. The jurisdictions would communicate the extent of gas incidents. They would share gas supplies for essential services where there is a gas supply shortfall that simultaneously affects several jurisdictions.

- **Access to the grid**

The Natural Gas Strategy adopted by the Commonwealth government in 1991 contained key objectives for gas reform that are still relevant today. These include competition through non-discriminatory open access to pipelines, the removal of regulatory barriers on interstate trade, more interstate interconnections, and a light-handed approach to regulation. In 1995, the states involved in the agreement joined the gas industry to set up a gas-reform task force that was to develop a national framework for grid access. "Natural Gas Pipelines Access Inter-Governmental Agreement" was adopted in November 1997. That document includes legislation on pipeline access. It contains an Access Code defining the rights and obligations of pipeline operators and users that apply for third-party access to natural gas transmission and distribution

---

61 Crocker K. (2002).

networks. To establish TPA in interstate trade, Gas Pipeline Access Act was adopted by the Commonwealth parliament in July 1998.

The Australian system resembles a negotiated TPA regime with a regulated commitment. This seems to be a good combination in a federal context, as it was acceptable to the various state governments.

AUSTRIA

- The seasonality of gas demand in Austria is important, due to the extensive use of gas in power plants for middle load.

- About a quarter of gas sales are on an interruptible basis.

- Russian gas, the major supply source, is received with a high load factor.

- Storage is the main load-management tool for matching seasonal demand with fairly constant imports. Austria has five underground gas storage sites and ample capacity, 2.8 bcm of working capacity, or 134 days of consumption.

- OMV is trying to develop Baumgarten as a trading hub.

GAS DEMAND

- **Share of gas in TPES:** 23% (2000)

Gas consumption reached 7.7 bcm in 2000. The industrial and power sectors are the major consumers of gas. The seasonality of gas demand in Austria is important, due to the extensive use of gas in power plants for middle load.

Figure 14: Austrian gas consumption by sector in 2000

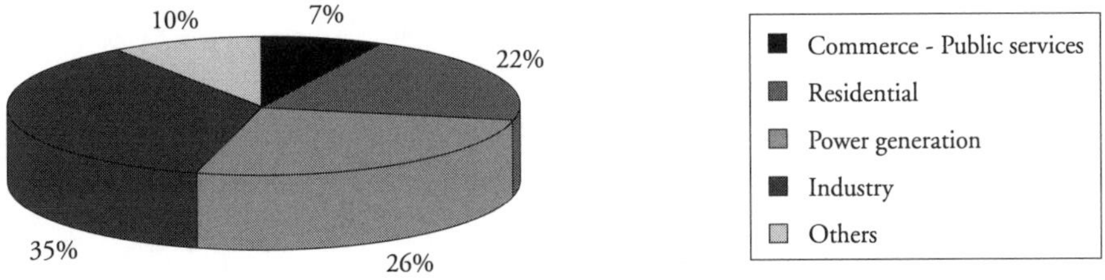

Legend:
- Commerce - Public services
- Residential
- Power generation
- Industry
- Others

- **Seasonality**

The ratio between gas sales in the peak and the lowest month of the year was about 3 to 1 in 2000. Peak daily sales by distribution companies were 41 mcm/d in December 2001, six times higher than the lowest daily sales in August 2001.

- Share of gas in the power-generation sector: 13% (2000)

Table 11: Multi-Fired Electricity Generating Capacity in Austria at 31 December 2000 (GW)

Solid/Liquid	0.28
Solids/Gas	1.21
Liquids/Gas	2.71
Liquids/Solids/Gas	0.12
Total Multi-fired	4.32
Total Capacity*	6.13

* From combustible fuels.

71% of total electricity generating capacity by fuel is multi-fired. The largest group of plants runs on natural gas.

- **Interruptibles and fuel switching**

Of total gas sales, 25% are sold on an interruptible basis.

Almost all gas-fired electricity plants are dual-or multi-fired.

GAS SUPPLY

- **Reserves:** 26 bcm   **R/P:** 14 years

- **Gas supply structure:** indigenous production 23%; imports 77%, of which:

    - Russia 63%

    - Germany 5%

    - Norway 9%

■ Supply swing

Gas imported mainly from Russia is imported with a high load factor. Production allows some swing, but it covers 23% of gas supply only.

Figure 15: Austrian monthly gas supply

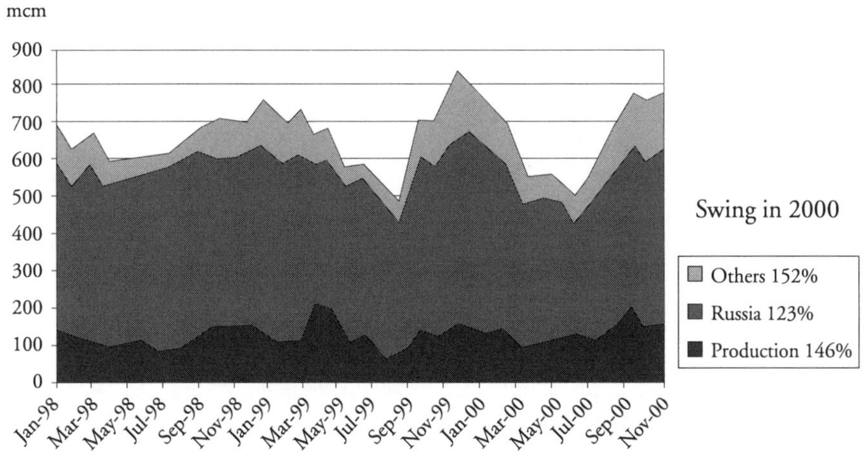

Others: Germany and Norway.

STORAGE

A volume of gas amounting to more than one-third of Austria's annual demand is held in five storage facilities, with a total working capacity of 2.8 bcm. This

Table 12: Underground gas storage in Austria

Name	Type	Operator/ Number	Working capacity (mcm)	Peak output (mcm/day)
Puchkirchen 1	Depleted gas field	RAG	50	0.5
Puchkirchen 2	Depleted gas field	RAG	450	4.5
Schoenkirchen/Reyersdorf	Depleted gas field	OMV	1,770	17.3
Speicher Vertrag 1	Depleted gas field	OMV	300	3.2
Thann	Depleted gas field	OMV	250	2.8
Total		**5**	**2,820**	**28.3**

represents 134 days of current average consumption. All the storage facilities are in depleted fields, mostly around Vienna.

■ **Stock changes**

Figure 16: Austrian load balancing

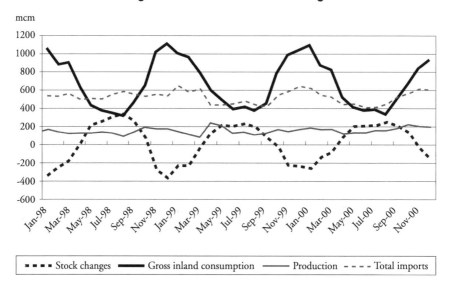

This graph highlights the very important role that storage plays in meeting seasonal requirements.

REGULATORY FRAMEWORK

■ **Gas security of supply**

Storage plays an important part in the emergency supply plan of the Austrian gas industry.

■ **Access to the transmission grid and storage**

Until the new Gas Law was passed, access to the transmission grid was negotiated on the basis of published general conditions and pricing structures. Since October 2002, access has been regulated. Access to storage is negotiated.

- **OMV's services**

OMV offers standard transport services, as well as a range of optional services including flexibility. OMV offers flexibility bids in its transportation service. Flexibility is individually negotiated with shippers.

OTHERS

Austria is an important transit country with some 25 bcm/year of Russian gas transiting Austrian territory en route to Italy, France, Germany, Hungary, Slovenia and Croatia.

OMV is developing "Gas Hub Baumgarten" (GHB), a trading hub which will include storage services in its portfolio. GHB is at the cross-roads of major lines. It has suitable technical infrastructure, including Eurostorage Baumgarten, located at the Austrian-Slovak border.

BELGIUM

- The seasonality of gas demand is relatively less pronounced in Belgium than in other EU countries.

- Belgium has the ability to interrupt large power generating plants that account for about a third of gas sales.

- The Interconnector offers new flexibility (spot purchases).

- The country plays an important transit role.

- The Dutch and Norwegian supplies provide a very large proportion of the total swing.

- There is little underground storage capacity due to unfavourable geological conditions.

- Belgium's hub, at Zeebrugge, offers great flexibility and reduces the need for storage.

- An LNG terminal offers additional flexibility.

GAS DEMAND

- **Share of gas in TPES:** 23% (2000)

Gas consumption reached 15.7 bcm in 2000. The power sector represents 23% of total demand, much of which is interruptible. The power sector uses high calorific gas. Since Distrigaz can cut off the power sector more easily than others, there tends to be a surplus of high calorific gas but a shortage of low calorific gas.

Figure 17: Belgian gas consumption by sector in 2000

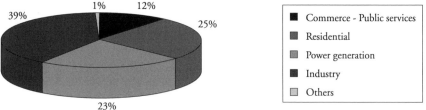

■	Commerce - Public services
■	Residential
■	Power generation
■	Industry
□	Others

■ Seasonality

The ratio between the peak and the lowest month of the year was 2.3 to 1 in 2000. Belgium's ability to interrupt large power generating plants that account for more than a third of gas sales allows it to cope with seasonal requirements in the residential sector.

■ Share of gas in the power-generation sector: 19% (2000)

Table 13: Multi-fired electricity generating capacity in Belgium at 31 December 2000 (GW)

Solid/Liquid	0.40
Solids/Gas	0.17
Liquids/Gas	4.37
Liquids/Solids/Gas	1.90
Total Multi-fired	6.84
Total Capacity*	8.55

* From combustible fuels.

80% of total electricity generating capacity by fuel is multi-fired and of this amount, plants running on natural gas have the largest share.

■ Interruptibles and fuel switching

According to a recent report by the Belgium Gas and Electricity Regulatory Authority[62], at least 30% of daily industrial demand on average is interruptible; 52% of industrial consumers have a contract for firm supplies, 21% for non-firm supplies and 27% for some combination non-firm and firm supplies. Among the interruptible customers, 76% can be interrupted by the transmission system operator (peak load balancing) and 24% at the request of the customer (price arbitrage); 62% of industrial customers with an interruptible contract can switch to an alternative fuel.

The CCGT plants can switch to gas oil.

---

62 CREG (2001).

GAS SUPPLY

- **Gas supply structure:** 15.7 bcm in 2000; indigenous production zero; imports 100%, of which:
  - Algerian LNG 28%
  - Norway 32%
  - Netherlands 35%
  - Spot (from Interconnector) 5%

The spot market is taking an increasing share. It was zero in 1998.

- **LNG share in gas supply:** 28%

- **Gas imports flexibility**[63]

Figure 18: Belgian monthly gas supply

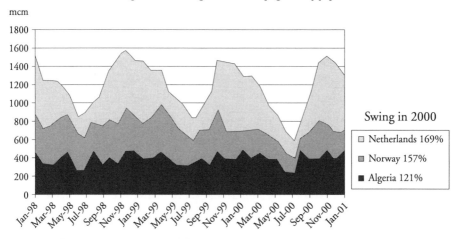

Does not include UK imports (UK imports flexibility amounted to 231% in 2000).

Dutch and Norwegian supplies provide a very large proportion of Belgium's total swing. Norwegian supply was relatively flat before 1999 (113% in 1998) and then swing begun increasing to 134% in 1999 and 157% in 2000. Algerian gas deliveries are nearly flat. The constraint on Algerian supplies is the availability of LNG tankers. Short-term contracts are also important. The graph, however, gives an idea of flexibility in long-term contracts.

---

63 Due to the confidential character of data, the government only partly publishes monthly data related to gas imports.

STORAGE

For geological reasons, storage is small in relation to demand; there is only one significant underground site. There is one LNG storage plant at Dudzele that is used for peak shaving.

Table 14: Underground gas storage in Belgium

Name	Type	Operator/ Number	Working capacity (mcm)	Peak output (mcm/day)
Anderlues	Mine	Distrigas	84	1.3
Dudzele	LNG	Distrigas	58	8.9
Loenhout	Aquifer	Distrigas	570	10.0
Total		**3**	**712**	**20.2**

- **LNG import terminals**

Table 15: Belgian LNG import terminals

Terminals	Storage tanks (1,000 cm of LNG)	Nominal capacity (mcm/day)	Start-up date
Zeebrugge	260	8.6	1987

- **Stock changes**

Figure 19: Belgian load balancing

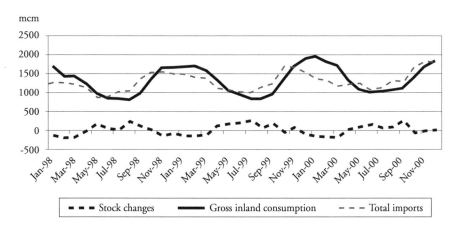

mcm

Stock changes — Gross inland consumption — Total imports

This graph shows that storage does not play a very important role in meeting seasonal fluctuations in gas demand. Flexibility in gas import contracts does play a major role. Purchases on the spot market have recently increased significantly.

REGULATORY FRAMEWORK

■ **Security of supply**

The "public service obligation" concept is important in Belgium. The network is designed to cover all firm clients with amounts based on those of the winter of 1962-63. Daily delivery must be guaranteed down to temperatures of minus 11°C, and sufficient gas must be stored to meet the volume requirement of a one-in-fifty winter. Hourly peaks are covered by line-pack.

■ **Access to transportation and storage**

The 1999 Gas Law established negotiated access based on published commercial conditions. However, the law was modified on July 18, 2001 and the regime has been changed to one of regulated third-party access for the entire gas system including storage and LNG terminals.

Belgium's limited storage capacity is entirely dedicated to the distribution market as Distrigas must cover the sector's demand under a public service obligation.

OTHERS

■ **Transit of gas**

Belgium is a key transit country. Its total transit capacity amounts to 48 bcm/year, or three times internal consumption. Norwegian, UK and Dutch gas can transit the country to France, Italy and Spain. Daily input and output figures through the transit pipelines are practically equal; so Belgium cannot exploit the transit of gas to help its own load-balancing. One exception to the rule can be found in the transit of gas to France. During the winter Belgium delivers to France less Dutch gas than it receives, effectively drawing gas from French storage. The pattern is reversed during the summer.

- **Zeebrugge Hub**

Zeebrugge, the landing site of the Zeepipe and the location of an LNG regasification terminal, has seen its role reinforced in 1998 with the opening of the Interconnector linking Zeebrugge to Bacton. The Interconnector has given birth to the first continental European hub, Zeebrugge Hub, where approximately 40 companies now regularly exchange gas. The liquidity at the Zeebrugge Hub is such that representative price indexes can now be established.

- **Distrigas services**

Virtual storage is available at published tariffs.

CANADA

- Canada has a large, fully deregulated gas market. Major production areas are in the west of the country, while the main consumers are in eastern Canada and the US.

- Gas sales to the residential sector in the peak month of the year are five times higher than in the lowest month. The residential and commercial sectors represent 33% of total sales.

- To cope with seasonal variations, Canada has developed huge underground gas storage, in both upstream and downstream regions. Total working capacity amounts to 17.2 bcm, or 19% of average annual gas consumption.

- Interruptible contracts are widely used, but precise estimates of the potential for switching between fuels are not available.

GAS DEMAND

- **Share of gas in TPES:** 30% (2000)

Canadian demand totalled 90.4 bcm in 2000, an increase of 6% from 1999. The industrial sector represents approximately 33% of total Canadian consumption. The other most important market in Canada is the "core sector" which includes space heating for residential and commercial buildings.

Figure 20: Canadian gas consumption by sector in 2000

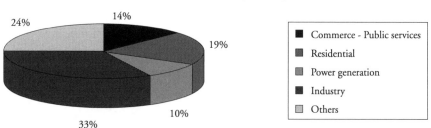

"Others" includes consumption by the natural gas industry itself, as pipeline compressor fuel.

■ Seasonality

The seasonality of gas demand is high, mainly because of weather patterns. The consumption profile of each market sector is important, as it defines the type of contracting practices and risk management the sector will pursue. The ratio between total gas sales in the peak and the lowest month of the year was 2.1 to 1 in 2000.

Figure 21: Canadian gas demand seasonality (2000)

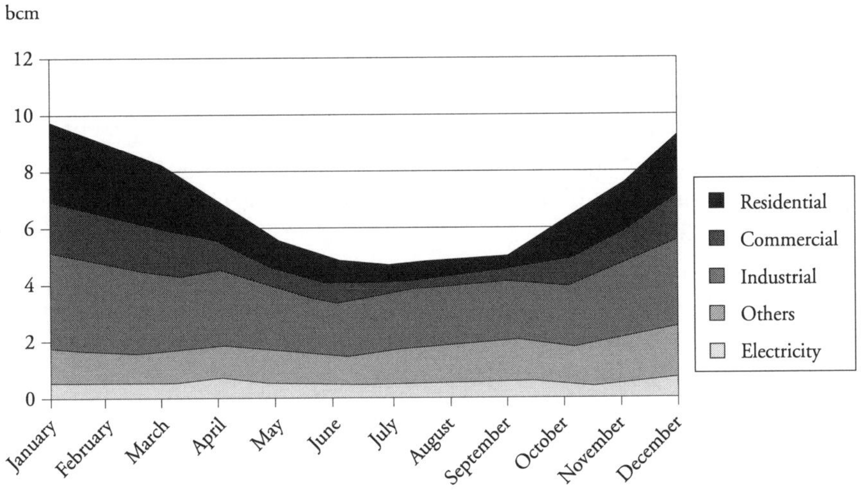

Source: Canadian Natural Gas, Market Review & Outlook, Natural Resources Canada, 2001.

■ Share of gas in the power-generation sector: 6% (2000)

Table 16: Multi-fired electricity generating capacity in Canada at 31 December 2000 (GW)

Solid/Liquid	0
Solids/Gas	1.63 (e)
Liquids/Gas	0.43 (e)
Liquids/Solids/Gas	0
Total Multi-fired	2.06
Total Capacity*	33.18

* From combustible fuels.

Only 6% of total electricity-generating capacity by fuel is multi-fired, with plants running on natural gas playing a marginal role. Most of electricity produced in Canada comes from hydro power.

■ **Interruptible contracts and fuel switching**

The use of interruptible contracts is widespread in Canada, but precise estimates of the potential for switching between fuels does not seem to exist. In 1995, the government indicated that 20% of industrial sales by local distribution companies, or 10% of total sales, were switchable.

The use of interruptible contracts varies widely by region; it is very low in areas where end-users are located near production sites. In eastern Canada, local distribution companies have low interruptible rates in order to encourage some customers to maintain multi-fuel capability. There are, however, few curtailments to interruptible customers, as curtailments generate loss of revenues.

GAS SUPPLY

■ **Reserves:** 1,728 bcm   **R/P:** 10 years

■ **Gas supply structure:** indigenous production 99%; imports 1%, from US.

■ **Gas production**

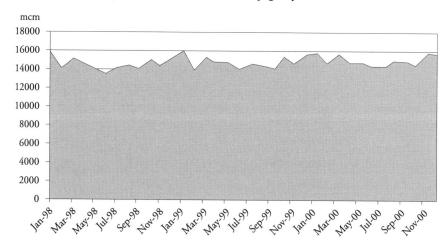

Figure 22: Canadian monthly gas production

Canadian production totalled 181.7 bcm in 2000, an increase of 3.1% over 1999. Most of the increase was due to the start-up of the Sable Offshore Energy project. Production is relatively flat throughout the year. Swing production was 106% in 2000.

■ **Gas exports**

Canadian natural gas exports to the US increased 7 bcm, or 7%, to reach 101.8 bcm in 2000. Higher exports to the Northeast US were mainly due to the start-up of the Maritimes & Northeast pipeline and Sable Offshore Energy project in January 2000. The swing factor of gas exports was 124% in 2000.

Figure 23: Canadian monthly gas exports to US

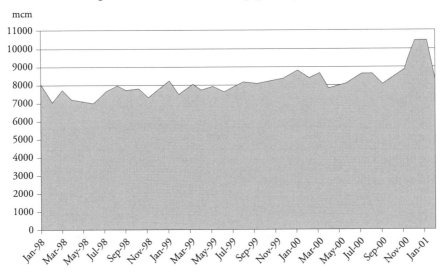

Pipeline capacity utilisation in Canada is generally high. According to Natural Resources Canada, existing export capacity was used at 90% load factor in 2000. The physical export capacity reached 12,100 mcfd (125 bcm/year or 342 mcm/day) when the Alliance project was completed in December 2000.

Right now total export capacity cannot be filled due to a lack of gas supply. Due to various constraints, capacity is seldom used at 100% load factors. In recent years, the best fill rate for total export capacity was about 95%.

Table 17: Export pipeline capacity (mcm/d)

	1999	2000	
	Year-end capacity	Increment	Year-end capacity
Huntingdon (Westcoast)	29.6	0.0	29.6
Huntingdon (User Pipes)	10.8	0.0	10.8
Kingsgate (Foothills/ANG)	73.1	0.0	73.1
Total to US West	113.5	0.0	113.5
Monchy (Foothills)	62.0	0.0	62.0
Emerson (TCPL)	37.0	0.0	37.0
Elmore (Alliance)	0.0	37.5	37.5
Miscellaneous (see note)	8.5	0.0	8.5
Total to US Midwest	107.5	37.5	145.0
Iroquois (TCPL)	25.0	0.2	25.2
Niagara Falls (TCPL)	23.9	0.0	23.9
Chippawa (TCPL)	14.2	0.0	14.2
St. Stephen (MNP)	10.2	0.0	10.2
E. Hereford (TCPL)	4.6	1.1	5.7
Cornwall (TCPL)	1.8	0.0	1.8
Napierville (TCPL)	1.7	0.0	1.7
Phillipsburg (TCPL)	1.4	0.0	1.4
Highwater (TCPL)	0.0	0.0	0.0
Total to US Northeast	82.8	1.4	84.2
Total capacity (export)	303.8	38.9	342.6

Notes: Year-end mcm/d capacity represents approximately the contracted daily volumes that could be delivered on the last day of the year. Capacity additions are generally completed on November 1. Miscellaneous (Midwest) includes nine export points with another 14 mcm/d of capacity. These export points are not intended to be used at high load factors, and so a lower number has been used in the table.

Source: Canadian Natural Gas Market Review & Outlook, Natural Resources Canada, 2001.

STORAGE

Total working storage capacity in Canada is 17.2 bcm, or 19% of gas consumption in 2000. The maximum daily sendout rate is 715 mcm. Upstream storage capacity is being expanded in the producing regions, as one of the responses to tighter supply and demand conditions.

Table 18: Underground gas storage in Canada

Operator	Type	Number	Working capacity (mcm)	Peak output (mcm/day)
AEC	Depleted gas field		2,691	56.6
BP Amoco and TransCanada	Depleted gas field		1,189	14.2
Andeerson Exploration	Depleted gas field		340	
B.C. Gas	LNG tanks		17	4
Consumers Gas	Depleted reef		2,720	48
GMI/Stogaz/Intragaz	Depleted gas field		57	7
Atco Gas	Salt cavern		3,116	482
Sabine Hub Services	Depleted gas field		992	21
Transgas	Salt cavern		907	14
Union Gas/Centra	Depleted reef		4,023	57
Unocal Canada	Depleted reef		1,133	11
Total			**17,184**	**714.8**

■ **Stock changes**

Figure 24: Canadian load balancing

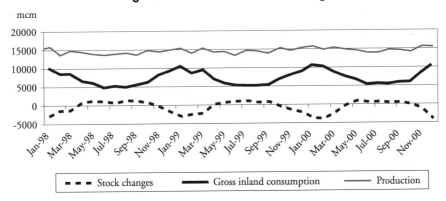

This graph shows the seasonality of gas demand and the role of gas storage in balancing supply and demand. Peak requirements are usually met by gas in storage.

The role of storage in Canada is essential. Since deregulation in 1985, excess deliverability has decreased. Many producers and marketers have increased the proportion of their total gas supply that is sold on a short-term or spot basis. This enables them not only to backstop their long-term commitments more readily, but also to take advantage of any short-term spikes.

For many years local distribution companies have used downstream storage facilities located near their markets as an efficient tool to manage their gas-supply portfolios and their customers' gas peak demand during the heating season. Downstream storage is increasingly used not only by local distribution companies, but by end-users, marketers and pipeline companies as a way of increasing the reliability of gas supplies.

REGULATORY FRAMEWORK

■ **Security of supply**

The underlying principle of the Canadian Government's approach to security of supply is to ensure a healthy and open natural gas market. Nevertheless, the Energy Preparedness Act and the Emergencies Act allow the Federal Government to allot, in an emergency, natural gas.

Energy security is the responsibility of the federal government. Physical energy security is not an issue in Canada because of the country's huge and diverse energy resource base. There are several major transmission pipelines, which transport gas produced in the west to the principal markets in the east. Vulnerability to supply disruption arising from long transmission pipelines is mitigated by duplicated lines. There is substantial upstream storage capacity in western Canada and downstream storage in eastern Canada.

The most vulnerable part of the system seems to be the TransCanada pipeline between Alberta and Ontario, which is the only direct east-west connection. The pipeline, however, is looped along its entire length, and disruptions could be countered by using storage in eastern Canada.

Currently, there are 24 pipeline interconnections between Canada and the US with a total annual maximum capacity of 343 mcm/d. To manage possible interruption in gas supplies, agreements have been made with large customers in Canada and the US, such as cogeneration plants, which are capable of burning other fuels than gas. Under the agreement, gas to those customers could be interrupted.

The Canadian government considers fuel choice to be a commercial matter. It no longer has policies or programmes to encourage multi-fired capacity.

- **Access to the transportation system**

Interprovincial transmission is regulated by the National Energy Board (NEB), which ensures that open non-discriminatory access is provided to all shippers on interprovincial gas pipelines. Interprovincial transportation rates, conditions of access and terms of service are regulated by the NEB. However, the board generally accepts the outcome of private negotiations for access to pipelines and imposes regulated rates only where a negotiated rate could not be agreed.

Local distribution companies are regulated at the provincial level by public utility commissions. These commissions regulate the rates charged by the companies for services, and authorise construction of transmission and distribution lines.

CZECH REPUBLIC

- The seasonality of gas demand is very high. The temperature-sensitive market accounts for more than 40% of gas sales.

- Supplies from Russia and Norway offer little swing.

- The Czech Republic relies heavily upon storage to cope with fluctuations in demand. The country has seven gas storage sites on its territory and ample capacity (2.1 bcm of working capacity, or about 80 days of consumption). The Czech Republic has also contracted for gas-storage capacity in the Slovak Republic and Germany.

GAS DEMAND

- **Share of gas in TPES:** 19% (2000)

Natural gas consumption increased by 80% in the Czech Republic during the last decade reaching 19% of TPES in 2000 (9.2 bcm). The temperature-sensitive market accounts for 43% of gas sales. Household consumption increased significantly during the last decade thanks to the extension of distribution networks, the replacement of town gas, the production of which ceased in 1996, and incentives to switch from coal to natural gas.

Figure 25: Czech gas consumption by sector in 2000

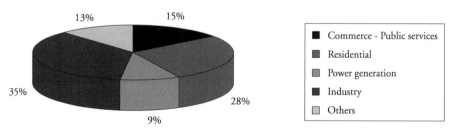

13%   15%

35%

9%   28%

- Commerce - Public services
- Residential
- Power generation
- Industry
- Others

- **Seasonality**

The ratio between gas sales in the peak and the low month of the year was 4.5 to 1 in 2000, which is very high compared with other European countries. The growth of space heating in the household and service sectors has increased the seasonal character of natural gas demand.

- **Share of gas in the power-generation sector:** 4% (2000)

There is no multi-fired power plants in the Czech Republic.

- **Interruptibles and fuel switching**

So far, interruptible contracts are not common in the Czech Republic. In the future, interruptible contracts with industrial users may provide enough flexibility to accommodate the seasonal fluctuations of gas demand and limit the need for new storage capacity.

GAS SUPPLY

- **Gas supply structure:** indigenous production 2%; imports 98%, of which:

  - Russia 77%
  - Norway 21%

The Czech Republic is a major transit country that plays a strategic role in Europe. Russian gas in transit through the Czech Republic represents nearly 25% of Western European imports.

- **Supply swing**

Russian and Norwegian supplies offer little swing (120% in 2000).

Figure 26: Czech monthly gas imports

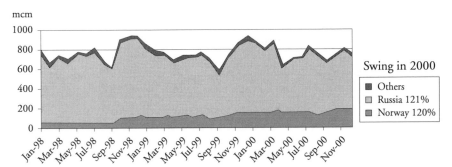

Others: German imports, i.e. less than 1% of the supply.

STORAGE

Six underground storage facilities (UGS) are owned by Transgas. UGS Uhrice is owned by Moravske Naftove Doly. All the underground facilities are operated by Transgas. Their total capacity is 2.1 bcm (2001 data).

Table 19: Underground gas storage in the Czech Republic

Name	Type	Operator/ Number	Working capacity (mcm)	Peak output (mcm/day)
Dolni Dunajovice	Depleted gas field	Transgas	700	12
Tvrdonice (Hrusky)	Depleted gas field	Transgas	487	7
Lobodice	Aquifer	Transgas	140	3
Stramberk	Depleted gas field	Transgas	435	6
Tranovice	Depleted gas field	Transgas	228	2.5
Haje	Depleted gas field	Transgas	57	6
Uhrice	Depleted gas field	Transgas	100	6
Total		**7**	**2,147**	**42.5**

The total storage capacity amounts to 23% of the yearly consumption. However, the supply situation has been tense during winter peak load periods and Transgas has had to lease additional storage capacity. Capacity under contract in gas storage sites at Láb in the Slovak Republic is 500 mcm, with peak load of 5 mcm/day, and at Rehden in Germany, 492 mcm, with peak load of 4.2 mcm/day.

■ **Stock changes**

Figure 27: Czech load balancing

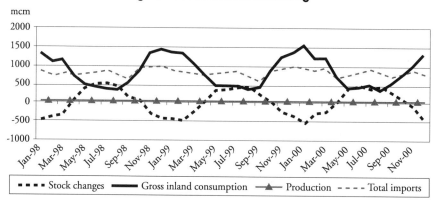

The graph highlights the important role of storage in meeting the Czech Republic's seasonal requirements.

REGULATORY FRAMEWORK

■ **Security of supply**

A new Energy Act was adopted in January 2001. The Act assigns power over the Czech market between the Ministry of Industry and Trade and a new Energy Regulatory Office which was established on 1 January 2001. The Energy Regulatory Office can impose an obligation to supply above the volume of energy given by licences in case of need or to offer access to gas transmission systems for the above-mentioned supply to gas companies.

Up to the opening of the Czech gas market on 1 January 2005, Transgas is responsible for secure and reliable supply. After market opening, Transgas, as transmission system operator, would be responsible for supply to all non-eligible customers.

Additional public service obligations that are given by the new Energy Act are as follows:

■ The transmission system operator, operators of distribution systems and natural gas storage are responsible for secure and reliable operation of their gas equipment.

■ The above mentioned gas operators are obliged to work out a contingency plan.

■ In case of an emergency, Transgas is obliged to steer the whole Czech gas network.

■ **Access to transportation and storage**

The responsibilities of the energy regulatory authority include setting up the framework for third-party access to the gas grid. The schedule for phasing in third-party access to gas starts with the largest users. The timetable goes as follows:

■ In 2005, 28% of the natural gas transmission system capacity will be opened to TPA.

■ In 2008, TPA will be extended to include 33% of natural gas transmission.

DENMARK

- Denmark has a relatively small gas market. Most of the gas is consumed by the power and combined heat and power generation sector.

- Seasonality of gas demand is less pronounced than in other European countries.

- Production, complemented by two gas storage facilities, offers sufficient swing to cover seasonal variations in gas demand.

GAS DEMAND

- **Share of gas in TPES:** 23% (2000)

Gas consumption reached 4.9 bcm in 2000. Half the gas is consumed by the power sector. There are five big power plants and numerous producers of combined heat and power and district heating. Consumption by the temperature-sensitive residential and commercial sectors, represents only 18% of total consumption.

Figure 28: Danish gas consumption by sector in 2000

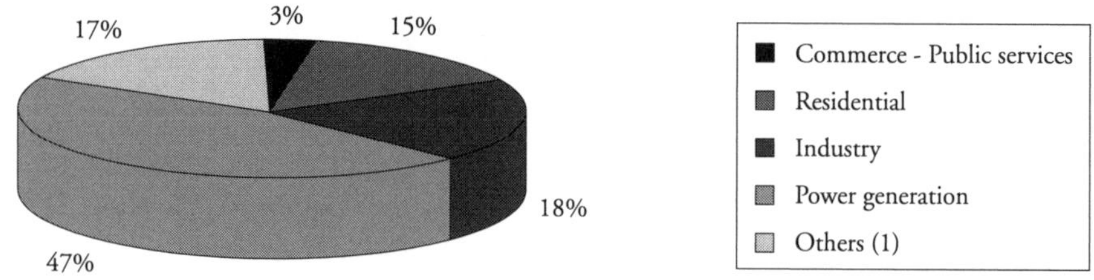

Legend:
- Commerce - Public services
- Residential
- Industry
- Power generation
- Others (1)

(1) The energy sector accounts for 87% of the "others" category.

- **Seasonality**

The ratio between gas sales in the peak and the lowest month of the year was 2.5 to 1 in 2000. The ratio between maximum and minimum monthly residential gas sales was 3 to 1 in 2000.

Figure 29: Danish seasonality in gas demand

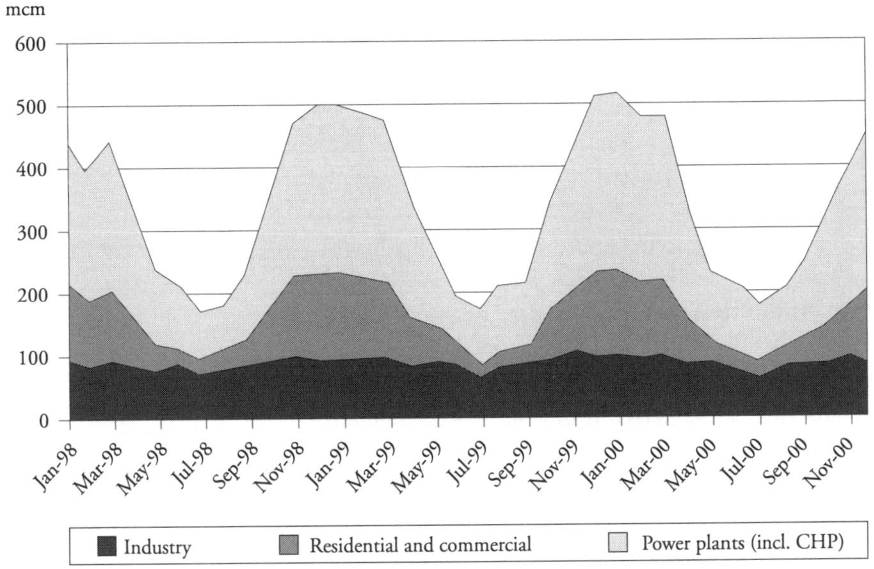

mcm

Industry | Residential and commercial | Power plants (incl. CHP)

Source: DONG.

■ Share of gas in the power-generation sector:: 24% (2000)

Table 20: Multi-fired electricity generating capacity in Denmark at 31 December 2000 (GW)

Solid/Liquid	5.49
Solids/Gas	0.25
Liquids/Gas	0.74
Liquids/Solids/Gas	0.36
Total Multi-fired	6.84
Total Capacity*	10.22

* From combustible fuels.

67% of total electricity generating capacity by fuel is multi-fired. In this share, plants running on natural gas have a marginal role.

■ Interruptibles and fuel switching

According to DONG, interruptible sales represented 33% of total gas sales in 2000. Half of sales to industry and 40% to the power sector are on an interruptible basis. Back-up fuels for interruptible customers are coal and oil.

GAS SUPPLY

- **Reserves:** 144 bcm   **R/P:** 18 years

- **Gas supply structure:** indigenous production 100%; imports zero

- **Production swing**

Denmark produced 8.2 bcm in 2000. The swing factor was 133%. 39% of the production is exported to Germany and Sweden.

Figure 30: Danish monthly gas production

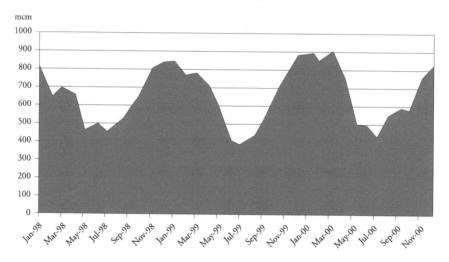

STORAGE

DONG has developed two underground storage facilities in aquifer and salt caverns that have a working capacity of 810 mcm, or 60 days of Danish consumption.

Table 21: Underground gas storage in Denmark

Name	Type	Operator	Working capacity (mcm)	Peak output (mcm/day)
Lille Torup	Salt cavity	Dansk Naturgas	410	13
Stenlille	Aquifer	Dansk Naturgas	400	11
Total		**2**	**810**	**24**

■ **Stock changes**

Figure 31: Danish load balancing

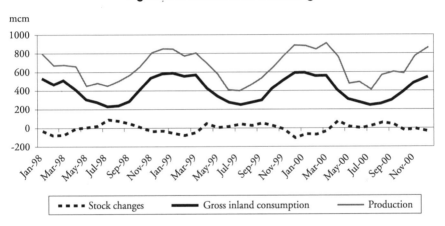

This graph shows that gas storage does not play an important role in balancing seasonal supply and demand. Denmark is able to modulate its production according to market needs.

REGULATORY FRAMEWORK

■ **Security of supply**

The Danish gas industry aims to guarantee supply to all non-interruptible customers:

■ even in the extreme case of a three-day supply failure at minus 14°C;

■ in the extreme case of a sixty-day supply failure at normal temperatures.

■ Access to transportation and storage

The Gas Act distinguishes between access to the transmission and distribution network and access to storage facilities. There is:

- regulated third-party access to transmission and distribution networks;
- negotiated access to storage facilities.

FINLAND

- Finland is a developing market supplied by Russian gas.

- Gas consumption is dominated by industry and power generation.

- The pipeline system covers only part of the country.

- Finland has no underground gas storage.

- Fuel-switching is an important tool in securing fuel supplies and managing seasonality.

- The opening of a secondary market, under the terms of the Natural Gas Market Act, provides a new flexibility tool to big customers.

GAS DEMAND

- **Share of gas in TPES:** 10% (2000)

Gas consumption reached 4.2 bcm in 2000. More than half of the gas is consumed by municipal district heating and related power production. Industrial consumers, mainly the pulp and paper industry, consume 27%.

Figure 32: Finnish gas consumption by sector in 2000

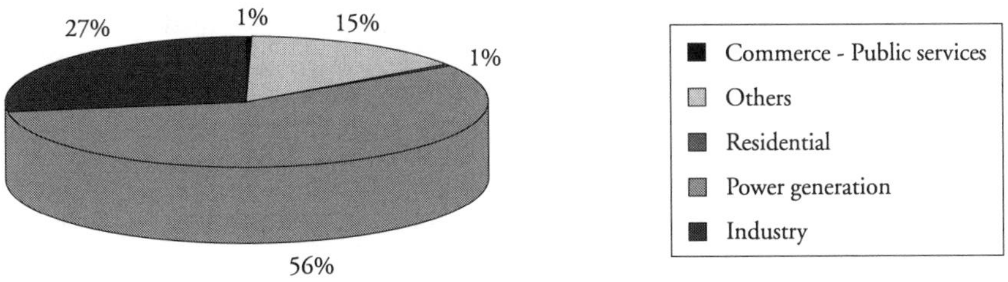

27%   1%   15%
1%

- Commerce - Public services
- Others
- Residential
- Power generation
- Industry

56%

- **Seasonality**

The ratio between gas sales in the peak and the lowest month of the year was about 2 to 1 in 2000.

- Share of gas in the power-generation sector: 14% (2000)

**Table 22: Multi-fired electricity generating capacity in Finland
at 31 December 2000 (GW)**

Solid/Liquid	5.23
Solids/Gas	0.25
Liquids/Gas	1.53
Liquids/Solids/Gas	2.13
Total Multi-fired	9.14
Total Capacity*	10.61

* From combustible fuels.

86% of total electricity generating capacity by fuel is multi-fired, with plants running on natural gas playing an important role.

- Interruptibles and fuel switching

Beyond formal interruptions, 90% of demand can easily switch to heavy or light fuel oil, approximately 3% can switch to LPG and the remaining 7% can be supplied through the pipeline network with a propane-air mixture.

GAS SUPPLY

- **Gas supply structure:** indigenous production 0%; imports 100%.

Natural gas deliveries to Finland are all from Russia. Although supplies have never been uninterrupted since the first gas deliveries in 1974, the priority for Finland now is to build a pipeline linking users to possible suppliers in the West.

In June 2000, a new parallel pipeline was connected and commissioned in the Karelian Isthmus in Russia. The parallel pipeline between Lappeenranta and Luumäki in Finland was completed and entered into commercial service in summer 2001.

- Supply swing

Gas is imported according to seasonal requirements. The swing in gas imports was 133% in 2000.

Figure 33: Finnish monthly gas imports

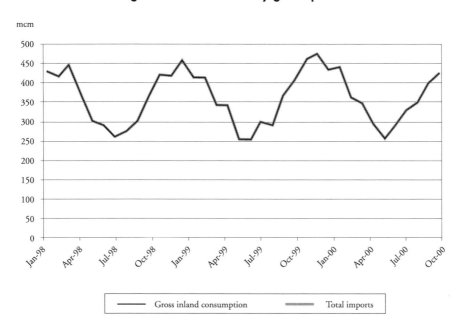

mcm

Legend: Gross inland consumption · Total imports

STORAGE

Finland's geology does not allow the construction of gas storage. Finland does store several back-up fuels.

REGULATORY FRAMEWORK

■ **Security of supply**

On the technical side, security of supply was improved with the completion of a second natural gas transmission line in the Karelian Isthmus.

To maintain security of natural gas supply, Gasum stores light and heavy fuel oils and propane, which can be mixed with air if necessary. The air-propane mixture can be pumped into the natural gas transmission line from the mixing plant located at Porvoo for those customers who are unable to use any other fuel.

In the event of supply disruption, natural gas can be replaced by back-up fuels. Non-industrial customers must have a back-up fuel capacity capable of bridging three months of natural gas consumption.

■ **Access to transportation**

According to the EU Gas Directive, Finland may be exempted from the obligation to open its natural gas networks until the Finnish gas network is directly connected to another EU country or until gas can be purchased from at least two external suppliers.

Finland's Natural Gas Market Act, which came into force on 1 August 2000, provided for the following changes from 1 January 2001.

■ Transmission and distribution companies (including Gasum) must separate their natural gas sales and network business in their accounts.

■ The Electricity Market Regulator was to be transformed into the Energy Market Authority with new responsibility in the gas sector.

■ There were to be new natural gas transmission and energy tariffs system. These entered into force at the beginning of 2001.

■ A "secondary market", consisting in trading of gas bought from Gasum but not used, was to be opened on 1 March 2001. Online trading of gas via the Internet did start in March 2001.

Eligible customers are not free to import gas from Russia (currently the only source). But retailers and large customers, those buying more than 5 mcm/year in 2000, may trade with each other any natural gas they have bought from Gasum and not used. These customers represent 90% of the market.

FRANCE

- France has a large temperature-sensitive market, representing more than half of gas sales.

- The seasonality of gas sales is high. Interruptible customers are not normally interrupted in winter.

- Suppliers offer little swing.

- France relies heavily upon storage to satisfy fluctuations in demand. It has 15 underground storage facilities, with working capacity of 10.5 bcm, or 95 days of consumption.

- The concept of "public service" is very important in France.

- Gaz de France offers "virtual storage" services through flexibility products.

GAS DEMAND

- **Share of gas in TPES:** 14% (2000)

Gas consumption reached 40.2 bcm in 2000. Sales to the residential-commercial sectors are rising steadily, with 52% of gas consumption in 2000. Sales to the industrial sector amounted to 41% of total consumption. The other sectors, in particular power generation (CHP), are marginal.

Figure 34: French gas consumption by sector in 2000

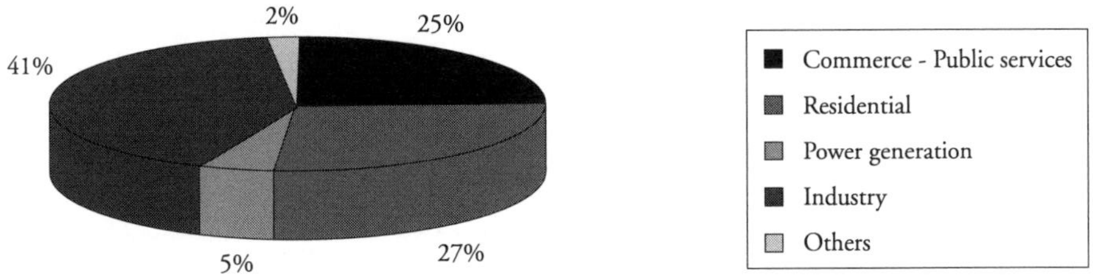

Legend:
- ■ Commerce - Public services
- ■ Residential
- ■ Power generation
- ■ Industry
- □ Others

2% 25% 41% 5% 27%

■ Seasonality

Figure 35: French gas demand seasonality

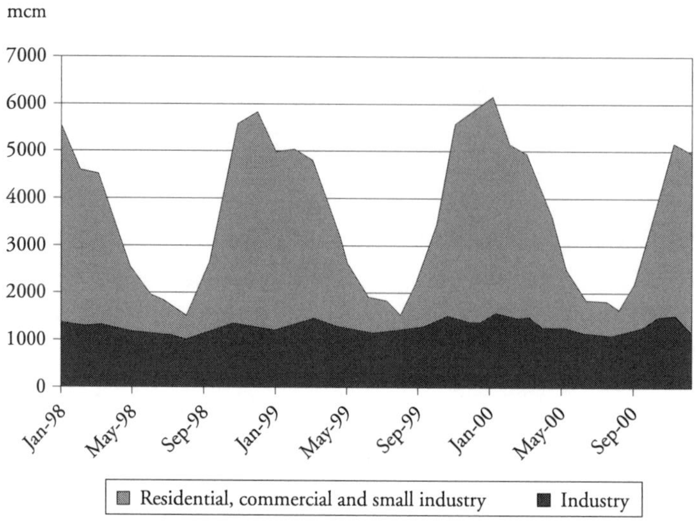

mcm

    ▨ Residential, commercial and small industry   ■ Industry

Source: Observatoire de l'Energie.

Minimum monthly demand in summer is about a quarter of monthly winter demand. The residential, commercial and small industry sectors account for almost all the seasonal variations in gas demand. In these sectors, sales are six times higher during the highest month than during the lowest month of the year.

■ **Share of gas in the power-generation sector:** 2% (2000)

Table 23: Multi-fired electricity generating capacity in France at 31 December 2000 (GW)

Solid/Liquid	8.62
Solids/Gas	0.69
Liquids/Gas	1.1
Liquids/Solids/Gas	0.92
Total Multi-fired	11.33
Total Capacity*	26.8

* From combustible fuels.

42% of total electricity-generating capacity by fuel is multi-fired, with plants running on natural gas playing a marginal role. The share of gas in the generation of electricity is marginal. Most electricity in France is generated from nuclear power.

- **Interruptibles and fuel switching**

Interruptible contracts make up approximately a third of the industrial market, but, they are not normally disrupted in winter. Disruptions would happen in case of gas supply interruption or of extreme cold weather conditions.

GAS SUPPLY

- **Reserves:** 8 bcm   **R/P:** 5 years

- **Gas supply structure:** indigenous production; 4%, imports 96%, of which:

  - Norway 28%

  - Russia 27%

  - Algeria 23%

  - Netherlands 12%

  - Nigeria 6%

- **LNG share in gas supply:** 23%

- **Supply swing**

Figure 36: French monthly gas supply

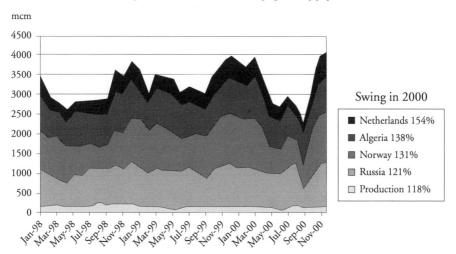

Does not include Nigerian imports.

The swing offered by external suppliers is a relatively flat 119% for total imports in 2000. Dutch gas offers less swing to France than it does to Belgium and Germany. First, because France does not border on the Netherlands, and it is desired that the transit pipeline through Belgium should be fully used. Second, because swap deals are made between Belgium and France, effectively using French storage, so that Belgium takes more than its contractual volumes during winter and less during summer.

STORAGE

France has 15 storage facilities, 12 in aquifers and 3 in salt cavities. The present working capacity is 10.5 bcm, or 95 days of average gas consumption. As France has only limited salt cavity storage it uses its aquifer storage in two modes: for traditional seasonal supply and also for peak supply.

Table 24: Underground gas storage in France

Name	Type	Operator/ Number	Working capacity (mcm)	Peak output (mcm/day)
Beynes Profond	Aquifer	GDF	310	9
Beynes Supérieur	Aquifer	GDF	210	4.5
Céré-la-ronde	Aquifer	GDF	320	3.7
Cerville-Velaine	Aquifer	GDF	640	4.8
Chéméry	Aquifer	GDF	3,430	42.5
Etrez	Salt cavity	GDF	450	20
Germiny-sous-Colombs	Aquifer	GDF	840	7
Gournay-sur-Aronde	Aquifer	GDF	670	17.5
Izaute	Aquifer	TFE	1400	1.8
Lussagnet	Aquifer	TFE	820	15
Manosque	Salt cavity	GDF	110	11
Saint-Clair-sur-Epte	Aquifer	GDF	410	4
Saint-Illiers	Aquifer	GDF	390	16
Soings-en-Sologne	Aquifer	GDF	240	9
Tersanne	Salt cavity	GDF	250	16.2
Total		**15**	**10,490**	**182**

- **LNG imports terminals**

Table 25: French LNG import terminals

Terminals	Storage tanks (1,000 cm of LNG)	Nominal capacity (mcm/day)	Start-up date
Fos-sur-Mer	150	22	1972
Montoir-de-Bretagne	360	31	1980

- **Stock changes**

Compared with other IEA countries the demand for load balancing in France is quite large. This is because the residential and commercial sectors hold a major share of total gas consumption.

Figure 37: French load balancing

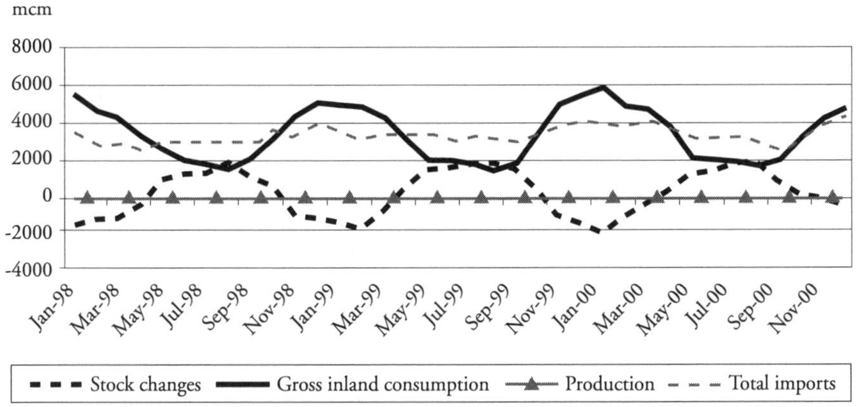

The graph clearly demonstrates the heavy reliance upon storage to accommodate seasonality, since suppliers offer very little swing.

REGULATORY FRAMEWORK

- **Security of supply**

France is dependent on imports for 97% of its gas supply, and 52% of these imports come from outside Europe. In this situation, France has opted to maintain strong security of supply, based largely on diversification of gas and energy supply. France has also a large gas storage capacity.

The gas system is designed to cope with any of the following situations:

- the coldest year that is statistically probable in a 50-year period;
- 1 in 50 years peak day demand.

- **Access to transportation, storage and LNG terminals**

A draft French law foresees a system of access based on published tariffs approved by the regulator. Negotiation would only be required in exceptional circumstances when specific conditions justify an individual contract.

The three major operators, GdF, CFM and GSO have published their tariffs under the temporary system and have adapted them to the regulator's observations.

GERMANY

- Germany is the second biggest European gas market after the United Kingdom.

- The temperature-sensitive market accounts for 40% of total demand and is growing.

- Germany enjoys considerable flexibility from indigenous production and imports from the Netherlands.

- The country has ample storage capacity. Its 39 underground storage facilities have 18.6 bcm working capacity, 75 days of consumption.

- Access to storage is part of an Associations Agreement. Storage terms have been in place since 30 June 2001.

- Most of the growth in European storage over the last decade, is concentrated in Germany.

- The country has the ambition of becoming a hub for European gas supplies.

GAS DEMAND

- **Share of gas in TPES:** 21% (2000)

German gas consumption reached 90.5 bcm in 2000. The largest share is in the temperature-sensitive residential and commercial sectors. Since 1995, the

Figure 38: German gas consumption by sector in 2000

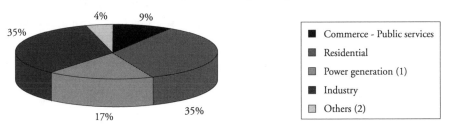

■	Commerce - Public services
■	Residential
▨	Power generation (1)
■	Industry
▨	Others (2)

(1) Includes combined heat and power.
(2) Includes district heating, energy sector consumption and transport.

industrial sector has taken about 40% of total consumption. The power sector represented 17% in 2000.

■ **Seasonality**

The ratio between gas sales in the peak and the lowest month of the year was about 3 to 1 in 2000.

■ **Share of gas in the power-generation sector: 9% (2000)**

Table 26: Multi-fired electricity generating capacity in Germany at 31 December 2000 (GW)

Solid/Liquid	9.46
Solids/Gas	2.58
Liquids/Gas	8.67
Liquids/Solids/Gas	6.90
Total multi-fired	27.61
Total capacity*	80.79

* From combustible fuels.

34% of total electricity generating capacity by fuel is multi-fired. Plants running on natural gas play an important role.

The share of natural gas in the power-generation fuel mix is continually increasing. In the new Länder, municipalities and industry have built new cogeneration plants based on natural gas for district heating, and gas is progressively replacing lignite in the existing district heating systems.

■ **Interruptible contracts and fuel switching**

Interruptible industrial contracts represent approximately a quarter of industrial consumption and they may be disrupted in winter. Gas companies have, however, been reluctant to sell interruptible contracts.

Approximately 70% of sales to the power sector are interruptible.

GAS SUPPLY

■ **Reserves: 264 bcm**   **R/P: 12 years**

- **Gas supply structure:** indigenous production 23%; imports 77%, of which:

  - Russia 35%

  - Netherlands 19%

  - Norway 19%

  - Denmark and spot 4%

- **Gas supply swing**

Figure 39: German monthly gas supply

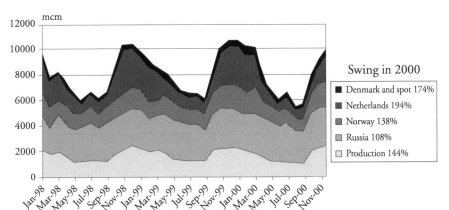

German produced 22 bcm of gas in 2000. Production is divided equally between sweet and sour gas. Capital intensive plants are needed to desulphurise sour gas (they also earn income from sulphur sales). As it is uneconomical to run them with a low load-factor, sour gas is produced virtually at the maximum technically possible rate. Sweet gas, on the other hand, is used as a major instrument of load management.

Most of the supply swing is provided by deliveries from the Netherlands (194% in 2000) and from German sweet production.

There are two principal qualities of gas: L-Gas and H-Gas. L-Gas has a high nitrogen content of 10-15%. Domestic production and imports from the Netherlands are mainly L-Gas. H-gas has less of inert gases, such as nitrogen. It may consist almost exclusively of methane, like Russian gas. Or it can contain a large share of higher hydrocarbons, as does Norwegian gas from associated fields or from gas condensate fields. L-Gas and H-Gas are handled separately by two transport, storage and distribution systems. The handling of two different qualities of gas further limits flexibility in supply and demand.

STORAGE

Due to the increase in household gas consumption and the consequent need for load balancing, German gas storage capacity has increased dramatically over the past ten years to 18.6 bcm in 2000. This represents 75 days of annual consumption. Underground storage is mainly owned by the regional transmission companies.

Table 27: Undergound gas storage in Germany

Name	Type	Operator/ Number	Working capacity (mcm)	Peak output (mcm/day)
Allmenhausen	Depleted gas field	CONT	40	0.8
Bad Lauschstädt b. Halle	Salt cavity	Verbundnetz Gas AG	806	22.3
Bad Lauschstädt b. Halle	Depleted gas field	Verbundnetz Gas AG	426	5.7
Berlin	Aquifer	Berliner Gaswerke AG	695	10.8
Bernburg	Salt cavity	Verbundnetz Gas AG	830	30
Bierwang b. München	Depleted gas field	Ruhrgas AG	1,300	28.8
Breitbrunn/ Eggestätt im Chiemgau	Depleted gas field	RWE-DEA AG Mobil EE GmbH Ruhrgas AG	550	6
Bremen-Lesum	Salt cavity	Stadtwerke Bremen AG	177	6.2
Buchholtz b. Postdam	Aquifer	Verbundnetz Gas AG	160	2.4
Burggraf-Bernsdorf bei Naumburg	Salt mine	Verbundnetz Gas AG	3	1
Dötlingen b. Oldenburg	Depleted gas field	BEB, Erdgas & Erdöl Gmbh	2,025	20.2
Empelde b. Hannover	Salt cavity	GHG GmbH (a)	146	7.2

Name	Type	Operator/ Number	Working capacity (mcm)	Peak output (mcm/day)
Epe (Ruhrgas)	Salt cavity	Ruhrgas AG	1,565	51
Epe (Thyssengas)	Salt cavity	Thyssengas AG	185	9.1
Eschenfelden bei Nürnberg	Aquifer	Ruhrgas AG EWAG	72	3.1
Etzel b. Wilhelmshaven	Salt cavity	IVG	534	31.4
Frankenthal b. Worms	Aquifer	Saar-Ferngas AG	60	2.4
Fronhofen	Depleted gas field	Deilmann EE GmbH for GVS	70	1.7
Hähnlein b. Darmstadt	Aquifer	Ruhrgas AG	80	2.4
Harsefeld b. Stadt	Salt cavity	BEB Erdgas & Erdöl GmbH	140	7.2
Huntorf I.d. Wesermarsch	Salt cavity	EWE AG	65	8.4
Inzenham - West bei Rosenheim	Depleted gas field	RWE-DEA AG for Ruhrgas	500	7.2
Kalle b. Bad Bentheim	Aquifer	VEW	315	9.6
Kiel-Rönne	Salt cavity	Stadtwerke Kiel AG & Schleswag	74	4.3
Kirchheiligen bei Mühlausen/Th	Depleted gas field		200	4.5
Kraak	Salt cavity		50	6
Krummhörn b. Emden	Salt cavity	Ruhrgas AG	116	6
Neuenhuntorf	Salt cavity	EWE AG	20	2.4
Nüttermoor b. Leer	Salt cavity	EWE AG	1,040	24
Rehden b. Diepholz	Depleted gas field	Wingas GmbH	4,200	57.6
Reitbrook b. Hamburg	Oil field with gas cap	Deilmann EE GmbH Hamburger Gaswerke	350	8.4
Sandhausen b. Heidelberg	Aquifer	Ruhrgas AG & Gasversor- gung Süddeutchland	30	1.1

Name	Type	Operator/ Number	Working capacity (mcm)	Peak output (mcm/day)
Schmidhausen b. München	Depleted gas field	Deilmann EE GmbH BMI Elwerath EE GmbH	150	3.6
Stassfurt	Salt cavity	VEW AG	68	5.3
Stockstadt b. Darmstadt	Depleted gas field, Aquifer	Ruhrgas AG	135	3.2
Uelsen	Depleted gas field	BEB EE GmbbH	660	7.4
Wolfersberg b. München	Depleted gas field	RWE-DEA AG for Bayerngas	400	5
Xanten am Niederrhein	Salt cavity	Thyssengas GmbH	193	6.7
Total		**39**	**18,556**	**425.2**

Germany has more storage than any other European country. It is the only European country that has substantial proportions of all three types of storage: depleted fields, aquifers, salt caverns. Most of the growth of European storage last decade took place in Germany, in line with increased gas consumption and with the development of new large sites, such as Rheden.

Germany has never explicitly aimed for a strategic reserve of gas. The main function of German storage is to cope with seasonal variations. Natural gas storage in Germany can be divided into two types: large volume storage for seasonal adjustment, which is mostly owned by the supra-regional companies, and peak storage for daily adjustment, which is owned by the regional or local gas companies.

■ **Stock changes**

Figure 40: German load balancing

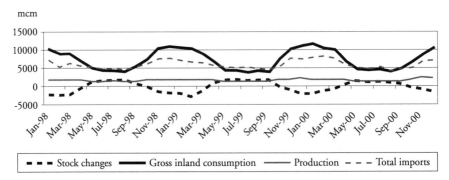

The graph shows that storage plays a complementary role in balancing supply and demand. Most of the supply swing is provided by deliveries from the Netherlands and by German sweet production.

REGULATORY FRAMEWORK

- **Security of supply**

In line with the light-handed regulation in Germany, the gas companies themselves are primarily responsible for ensuring the security of gas supplies. There are no standard criteria for security as there are in other European countries, such as to meet 1-in-20 or 1-in-50 winters. In developing their storage and transportation systems, German companies usually base their calculations on the winter 1962/63, which was the coldest in sixty years.

- **Access to transmission and storage**

Germany has opted for negotiated access to transmission and storage. An Associations Agreement (Verbändervereinbarung; acronym: VV), between two industrial-user associations and two gas-industry associations defines the basis for freely negotiated contracts. Access conditions to commercial storage facilities were defined in a second Associations Agreement (VV2), which refines the first. A third agreement, in May 2002, further defines access conditions.

In case of conflict over access or access conditions and tariffs, the dispute may be brought before the federal anti-trust authority (Bundeskartellamt).

Companies are required to publish the main commercial conditions for access to their network and storage. Since the introduction of negotiated access several dozens of transportation agreements have been concluded. Ruhrgas, Wingas, BEB, Thyssengas and VNG, the five biggest holders of gas storage capacity in Germany, have published terms and fees for third-party access to their facilities.

Ruhrgas offers storage only as part of a package with transportation to specific injection and withdrawal points, a service which it describes as "virtual storage." Most other providers, including Wingas and BEB, sell storage separately from transportation.

■ **Trading instruments, hubs**

A gas-trading hub is being developed at Bunde in Germany. Bunde is close to
the delivery points for Norwegian gas in Emden and Dornum, and close to the
main supply point for Dutch gas in Oude Stadenzijl. It is also close to German
domestic production as well as to the large storage facilities at Etzel, Bunde and
Rheden and to the north-south and east-west German pipelines.

GREECE

- Greece is currently establishing its gas market. Gas consumption reached 2.1 bcm in 2000, most of it used by power plants. Seasonal variations in gas demand are limited.

- Greece has no indigenous production. Imports from Russia and Algeria offer little swing.

- Greece has no underground gas storage but the Revithousa LNG regasification plant has storage tanks.

- As an emergent market, flexibility is not yet an issue in Greece. With the expected increase in gas demand, flexibility tools will have to be developed, such as an interconnection with the European gas grid and underground gas storage.

GAS DEMAND

- **Share of gas in TPES:** 6% (2000)

The Greek gas market is a young one. Natural gas was introduced in 1997 with the first imports from Russia. Gas consumption reached 2.1 bcm in 2000. As in most newly-developed markets, gas is primarily used in the power sector and industry.

- **Seasonality**

Figure 41: Greek gas consumption by sector in 2000

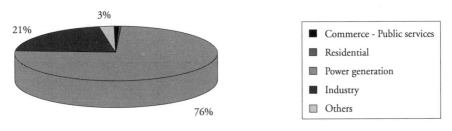

The ratio between gas sales in the peak and the lowest month of the year was 1.7 to 1 in 2000.

- **Share of gas in the power-generation sector:** 11% (2000)

There are no multi-fired power plants in Greece. About half of the existing generating capacity is based on coal and 30% on nuclear.

GAS SUPPLY

- **Reserves:** 1 bcm   **R/P:** 28 years
- **Gas supply structure:** indigenous production 2%; imports 98%, of which:
  - Russia 74%
  - Algeria 24%
- **LNG share in gas supply:** 24%
- **Supply swing**

Figure 42: Greek monthly gas imports

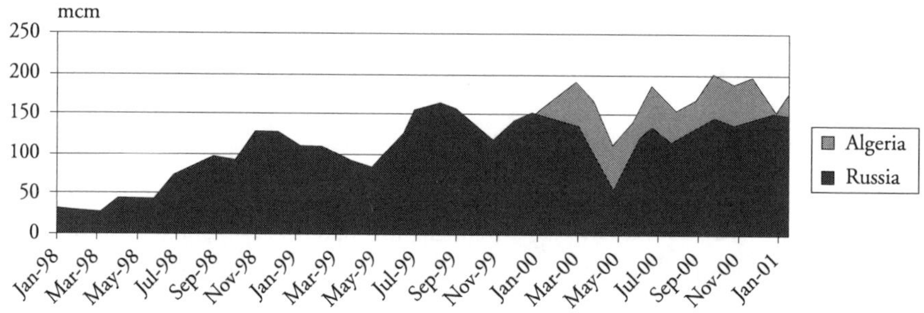

Russian deliveries started in 1997 and continue to rise. Algerian LNG deliveries to the Revithousa LNG regasification terminal started in February 2000. The supply swing from these deliveries is not yet a relevant factor.

STORAGE

Greece has not yet built any underground gas storage. However, there is a project to use a depleted gas field in South Kavala.

- **LNG import terminals**

The LNG regasification plant at Revithousa has storage capacity of 75,000 cm of LNG.

Table 28: Greece LNG import terminals

Terminals	Storage tanks (1,000 cm of LNG)	Nominal capacity (mcm/day)	Start-up date
Revithousa Islet	75	5.4	1999

■ **Stock changes**

Figure 43: Greek load balancing

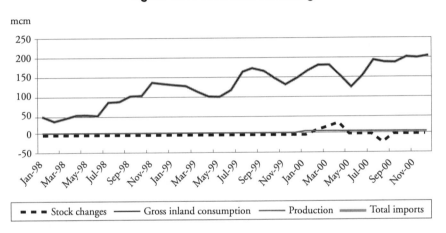

Seasonal variations in gas demand are limited, as gas is mainly used in power plants for base load. The Revithousa LNG terminal has storage tanks that can be used to smooth demand variations.

REGULATORY FRAMEWORK

■ **Access to the transportation grid**

Greece is an emerging market, not directly connected to the interconnected system of any other EU member state. Russia has a market share close to 75%. As a result, Greece did not have to apply the provisions of the EU Gas Directive in 2000. The opening of the Greek gas market is scheduled for 2006.

HUNGARY

- The seasonality of gas demand in Hungary is high. The temperature-sensitive market accounts for nearly half of gas sales.

- Indigenous production offers some swing, but domestic resources are declining and Hungary is becoming increasingly dependent on Russian gas imports.

- Hungary has five underground storage facilities with a working capacity of 3.6 bcm, or 110 days of Hungarian gas consumption.

- Hungary relies heavily upon storage to satisfy fluctuations in demand and is developing storage services for third parties.

GAS DEMAND

- **Share of gas in TPES:** 39% (2000)

Gas consumption reached 12 bcm in 2000. After the collapse of the old political and economic system in 1990, gas demand shrank, especially in the industrial sector. The share of the industrial sector in total gas consumption fell from 42% in 1990 to only 18% in 2000. On the other hand, the share of the commercial and residential sectors has increased. As a result gas consumption is highly seasonal.

Figure 44: Hungarian gas consumption by sector in 2000

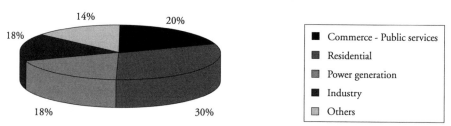

- **Seasonality**

The ratio between gas sales in the peak and the lowest month of the year was 4.2 to 1 in 2000, which is very high compared with other European countries.

■ Share of gas in the power generation sector: 19% (2000)

Table 29: Multi-fired electricity generating capacity in Hungary at 31 December 2000 (GW)

Solid/Liquid	0
Solids/Gas	0
Liquids/Gas	3.78
Liquids/Solids/Gas	0.20
Total multi-fired	3.98
Total capacity*	6.41

* from combustible fuels.

62% of total electricity-generating capacity by fuel is multi-fired, with plants running on natural gas holding the biggest share.

■ **Interruptibles and fuel switching**

Interruptible contracts are widely used for industrial customers that have fuel-switching capabilities.

Gas-fired plants are multi-fired and can burn fuel-oil when necessary.

GAS SUPPLY

■ **Reserves:** 26 bcm   **R/P:** 8 years

■ **Gas supply structure:** indigenous production 26%; imports 74%, of which:

- Russia 65%

- Germany* 6%

- France* 3%

* The gas received from Germany and France is, in fact, from Russia, but the contracts are signed with Ruhrgas and Gaz de France.

■ **Supply swing**

Domestic production offers some swing (136% in 2000), but it covers only 26% of total supply. Imports come from Russia with a high load factor; they offer very little flexibility.

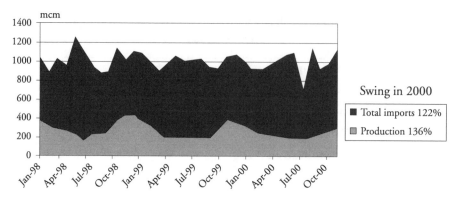

Figure 45: Hungarian monthly gas supply

Swing in 2000
■ Total imports 122%
□ Production 136%

STORAGE

Hungary has five underground storage facilities with a working capacity of 3.6 bcm, representing one-third of annual consumption. All the storage facilities are in depleted fields.

Table 30: Underground gas storage in Hungary

Name	Type	Operator/ Number	Working capacity (mcm)	Peak output (mcm/day)
Zana-Nord	Depleted gas field	MOL	1,300	18
Hajduszoboszlo	Depleted gas field	MOL	1,590	20.1
Pusztaederics	Depleted gas field	MOL	330	2.88
Kardoskút	Depleted gas field	MOL	240	3.4
Maros-1	Depleted gas field	MOL	150	2.2
Total		**5**	**3,610**	**46.58**

Source: IEA, MOL.

■ **Stock changes**

The graph below highlights the very important role of storage in meeting seasonal requirements. Peak winter demand is met by indigenous production and storage.

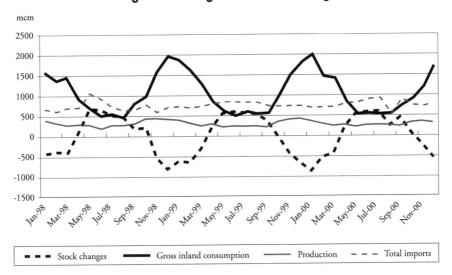

Figure 46: Hungarian load balancing

mcm

Legend: Stock changes — Gross inland consumption — Production — Total imports

REGULATORY FRAMEWORK

■ Security of supply

The main focus of the Hungarian government's energy policy is supply security. Increasing gas storage capacity plays an important role in its plans. Hungary has a competitive advantage because it has abundant depleted gas fields. The country has an explicit policy of responding to its declining domestic gas production and its increasing dependency on Russia through strategic storage.

Interconnection of the gas system with the European gas network has been accomplished with a pipeline linking Hungary and Austria (HAG) and long-term contracts have been signed with Ruhrgas and Gaz de France to diversify gas supplies. The HAG pipeline provides a link to the Western European gas grid. It allows both cross-border trade and a diversification of supply routes. The number and volume of contracts with the HAG are increasing. There is now virtual diversification of gas supply through swaps contracts with Western suppliers.

■ Access to transportation and storage

The Hungarian Government intends to implement the European Union Gas Directive (98/30/EC) by the time the country is a full member of the EU.

179

IRELAND

- The use of gas for base-load power generation represents more than 50% of Ireland's total demand.

- As a result, the seasonality of gas demand is less pronounced than in other European countries.

- The Irish market is interconnected to the UK market. Three fourths of the gas consumed is imported from UK. The balance is produced offshore Ireland.

- As there is sufficient capacity in the Interconnectors, Ireland has access to all of the flexibility mechanisms available on the UK gas market.

- Ireland has no underground gas storage of its own, though the South West Kinsale Field has been developed to help load-balancing.

GAS DEMAND

- **Share of gas in TPES:** 23% (2000)

Gas consumption reached 4 bcm in 2000. The power generation and industrial sectors are the principal gas consumers, representing 52% and 25% of total gas sales.

Figure 47: Irish gas consumption by sector in 2000

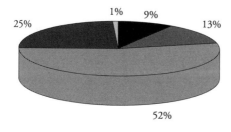
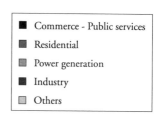

■ **Seasonality**

The ratio between gas sales in the peak and the lowest month of the year was 1.3 to 1 in 2000. The ratio between the peak day and the summertime low is approximately 6 to 1 in the residential and small commercial/industrial sectors. The ratio of the predicted peak month to the lowest month for these sectors is approximately 3.5 to 1.

■ **Share of gas in the power-generation sector:** 39% (2000)

Table 31: Multi-fired electricity generating capacity in Ireland at 31 December 2000 (GW)

Solid/Liquid	-
Solids/Gas	-
Liquids/Gas	1.60
Liquids/Solids/Gas	-
Total multi-fired	1.60
Total capacity*	4.07

* From combustible fuels.

39% of total electricity generating capacity by fuel is multi-fired by gas and liquids.

The recent high rate of growth in the Irish economy has led to a large increase in electricity demand. It is likely that incremental power-generation demand will be met through combined-cycle gas turbine plants for the foreseeable future. The power-generation share of total gas demand is expected to grow over time.

■ **Interruptibles and fuel switching**

The large power generation sites at Poolbeg, Aghada and North Wall have dual-fuel capability. Switching to heavy fuel-oil occurs when prices favour the burning of oil and also when natural gas supplies are restricted.

GAS SUPPLY

■ **Reserves:** 35 bcm   **R/P:** 9 years

■ **Gas supply structure:** indigenous production 29%; imports from UK 71%.

The Kinsale Head and Ballycotton fields are the current sources of indigenous supply. Both fields are in depletion and are expected to cease production around 2005. A new gasfield off the west coast, the Corrib Field, has been declared commercial. It is planned to bring this gas ashore starting from 2003/2004.

Irish imports represent around 4% of overall UK demand. Due to the relatively flat nature of Irish demand, the Irish market places a negligible burden on UK flexibility.

- **Supply swing**

Figure 48: Irish monthly gas supply

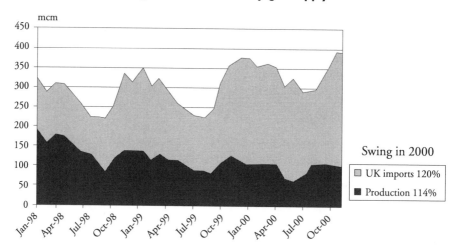

Bord Gáis has signed a re-profiling agreement with Marathon, the operator of the Kinsale Head Facilities Gas Field. South West Kinsale, a small satellite of the Kinsale Head Field, will be used to reinject gas in summer and withdraw it as needed in winter. At this early stage its primary function will be to reduce gas price differentials between summer and winter and to protect against price spikes in the spot market rather than to serve as a source of peak management. The re-profiling service has commenced and will be operational until 2005.

Bord Gáis's swing requirements are met from the UK, where a liquid spot market is in operation. Gas may be purchased either within-day or day-ahead to meet demand swings and to ensure daily balancing of the system.

STORAGE

There is no natural gas storage in Ireland, other than what will be provided by Marathon from the South West Kinsale gas field.

- **Stock changes**

Figure 49: Irish load balancing

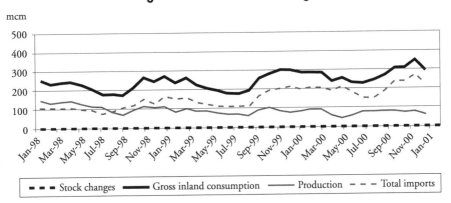

Seasonal variations in demand are mostly met by swing in imports.

REGULATORY FRAMEWORK

- **Security of supply**

The overall transmission system is designed to meet a 1-in-50 winter based on historic weather patterns.

Ireland has two sources of supplies: UK North Sea gas and indigenous production. The key to Irish security of supply till now is that either source could supply the non-interruptible gas market if there were a supply interruption in the other. However, demand now exceeds the delivery capacity of the Kinsale Head field. Starting from late 2002, a new interconnector with the UK will provide supply security for the older one. Its potential capacity is larger than that of the existing system. Either interconnector will be able to cover for a supply disruption in the other. From 2003, gas from the Corrib gas field will further enhance the system's security of supply.

- **Access to transmission**

In 1995, third-party access legislation was enacted in Ireland for sites using more than 25 mcm per year, or about 70% of the market. Ireland has opted for regulated third-party access. A division of the Department of Public Enterprise acts as regulator. Legislation going now before parliament will extend the role of the existing Commission for Electricity Regulation and will rename it as Commission for Energy Regulation, covering both gas and electricity. This legislation also provides for the lowering of the third-party access threshold to 2 mcm per year.

ITALY

- Italy has a highly developed gas market. Natural gas makes up 34% of total primary energy supply.

- Gas sales to the residential and commercial sectors account for 35% of the market.

- Seasonal variations in sales are extremely wide, with 62% of gas sales in the six winter months. There are few interruptible contracts.

- Italian producers and external suppliers offer very little swing, except the Netherlands.

- Italy is heavily dependent upon storage to cover fluctuations in gas demand. Italy has 10 underground storage sites, all in depleted reservoirs. The working volume is 12.7 bcm, 66 days of consumption.

- Italy holds substantial strategic stocks. For security of supply considerations, gas importers must store gas in a quantity equal to 10% of their total imports from non-EU countries. Enough back-up storage must be kept to meet the expected peak demand.

GAS DEMAND

- **Share of gas in TPES:** 34% (2000)

Gas consumption reached 70.7 bcm in 2000, having developed strongly over the past ten years, from 47.6 bcm in 1990. Since 1990, the share of the residential and commercial sectors in total consumption has stayed around 35-40%, whilst that of power generation has risen from 20% to 32%.

Figure 50: Italian gas consumption by sector in 2000

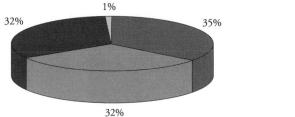
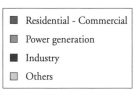

Residential - Commercial
Power generation
Industry
Others

■ Seasonality

Seasonal variations in sales are extremely wide, with 62% of gas sales taking place during the six coldest months. Sales to industry and power plants are flat throughout the year, but sales to the residential and commercial sectors are highly dependent on temperature. The overall ratio between gas sales in the peak and the lowest month of the year was 2.7 to 1 in 2000.

Figure 51: Italian seasonal gas demand

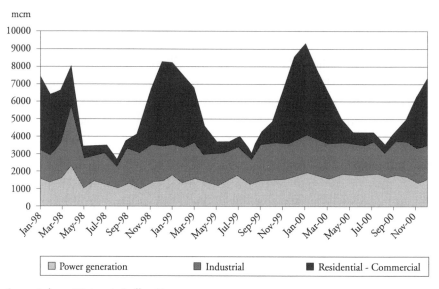

Source: Industry Ministry, in Staffetta News.

■ **Share of gas in the power-generation sector:** 38% (2000)

Table 32: Multi-fired electricity generating capacity in Italy at 31 December 2000 (GW)

Solid/Liquid	7.69
Solids/Gas	0
Liquids/Gas	20.36
Liquids/Solids/Gas	4.85
Total multi-fired	32.90
Total capacity*	54.03

* from combustible fuels. Public utilities only (data from autoproducers are not available).

62% of total electricity-generating capacity by fuel is multi-fired, with plants running on natural gas playing the biggest role.

■ **Interruptible contracts and fuel switching**

There is a trend in Italy towards firm gas contracts. Only 9% of sales are interruptible. Interruptible contracts exist only in the industrial sector. The conclusion of interruptible contracts is conditional upon the customer's having back-up facilities. 100% of interruptible industrial customers can use heavy fuel-oil as a substitute fuel.

Most gas-fired electricity generation plants are multi-fired, despite the fact that gas sales for power generation are not formally interruptible.

At the end of 2001 the share of interruptible capacity booked by shippers for the years 2001-2002 was around 9%. As the system now provides incentives to transmission companies to increase the volume transported, an increase in interruptible contracts can be expected in the coming years.

GAS SUPPLY

■ **Gas reserves:** 199 bcm   **R/P:** 13 years

■ **Gas supply structure:** indigenous production 22%; imports 78%, of which:

  - Algeria 38%

  - Russia 29%

  - Netherlands 8%

  - Nigeria 3%

■ **LNG share in gas supply:** 4%

■ **Supply swing**

Italian producers and external suppliers offer very little swing. The very flat swing in production can be explained by technical and economic factors. Before liberalisation, ENI/Agip treated the combination of their producing fields and storage facilities as a single system, which they sought to optimise. The gas-optimisation system is now regulated by the network and storage codes. Before the codes' approval, it was regulated by guidelines issued by the Ministry of the Productive Activities (MAP).

Figure 52: Italian monthly gas supply

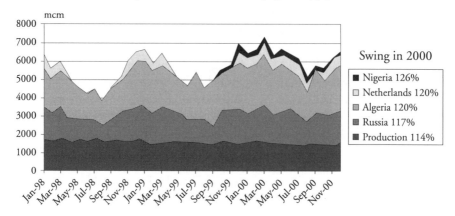

Except for the Netherlands, all of Italy's suppliers offer a high load factor. The Dutch contract is particularly flexible as all the gas received from the Netherlands comes during cold months.

STORAGE

There are ten storage facilities, all located in depleted gas reservoirs, mainly in Northern Italy. These facilities have a total working volume of 12.7 bcm, or 18% of total sales and 50% of sales to the residential and commercial sectors.

Table 33: Underground gas storage in Italy

Name	Type	Operator/ Number	Working capacity (mcm)	Peak output (mcm/day)
Brugherio	Depleted gas field	STOGIT	300	11
Cellino	Depleted gas field	EDISON TGS	110	na
Conegliano	Depleted gas field	EDISON TGS	545	na
Cortemaggiore	Depleted gas field	STOGIT	960	18
Minerbio	Depleted gas field	STOGIT	2,360	65
Ripalta	Depleted gas field	STOGIT	1,580	20
Sabbioncello	Depleted gas field	STOGIT	847	22.5
S.Salvo	Depleted gas field	STOGIT	2,895	44
Sergnano	Depleted gas field	STOGIT	2,000	55
Settala	Depleted gas field	STOGIT	1,150	31.5
Total		**10**	**12,747**	**267**

The facilities were until recently owned and operated by ENI/AGIP and EDISON. However, in 2001, ENI's board approved a plan to spin off its gas transportation unit, Snam, as well as its storage assets and service. Two new companies have been created: SNAM Rete Gas for transportation, regasification and dispatching, and Stoccaggi Gas Italia (STOGIT) for gas storage. Approximately 40% of SNAM Rete Gas has been privatised.

Italy has a very large volume of strategic gas storage. Historically, the strategic volume corresponded to six months-supply from the main supplier (Algeria in the 1980s). There has never been any law, obligation or government directive on the issue. ENI set the strategic reserve volume on its own initiative. Progressively lower ratios have been accepted, because supply has been extensively diversified. Current strategic volumes, 5.1 bcm in 2000, can cover a disruption for three-months on import pipeline or a very cold winter conditions.

As consumption increased from 47 bcm in 1990 to 71 bcm in 2000, so the amount required to cover seasonal fluctuations has increased too. New storage projects will boost storage capacity to 19.7 bcm and daily output to 300 mcm.

■ LNG import terminals

Table 34: Italian LNG import terminals

Terminals	Storage tanks (1,000 cm of LNG)	Nominal capacity (mcm/day)	Start-up date
Panigaglia	100	10	1971

■ New LNG import terminals

There are many projects for new LNG receiving terminals under consideration. These could eventually improve supply flexibility. New projects include: Rivigo (offshore), Taranto (Puglia), Vado Ligure (Liguria), Brindisi (Puglia), Muggia - Trieste (Friuli Venezia Giulia), Corigliano (Calabria), Lamezia Terme (Calabria), and Rosignano Marittimo (Toscana).

■ Stock changes

The graph below shows the important role played by gas storage in Italy in balancing gas demand and supply.

Figure 53: Italian load balancing

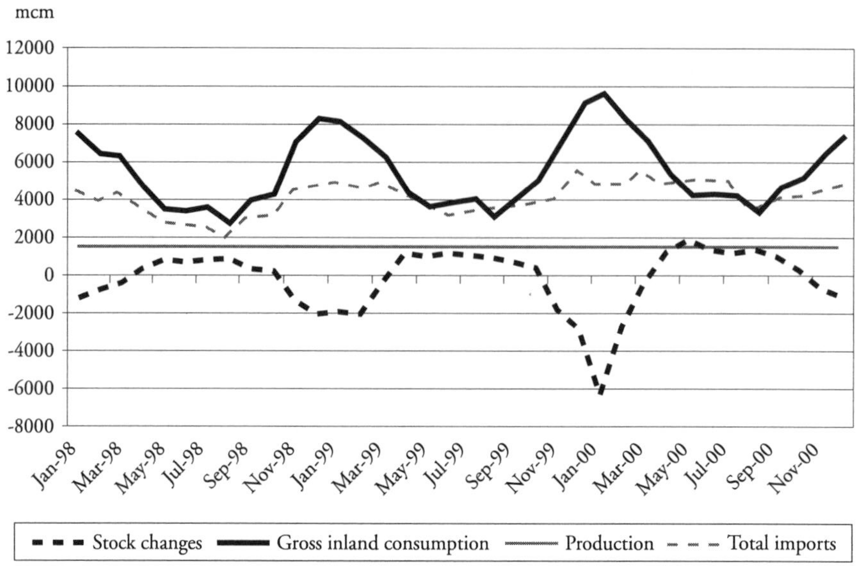

Legend: Stock changes · Gross inland consumption · Production · Total imports

REGULATORY FRAMEWORK

■ **Security of supply**

Transmission and storage systems have been designed to meet gas demand on the coldest winter day in 1-in-20 winter and gas interruption of six months due to disruption of the largest supplier, against the backdrop of a standard cold winter.

According to the decree regulating natural gas imports of March 2001, gas imports from non-EU countries require the approval of the Ministry of Productive Activities. In addition to meeting technical and financial requirements, the importer needs to demonstrate:

■ Availability of strategic storage equivalent to 10% of the volume annually imported from any non-EU source and, at the end of the peak season, to 50% of the average expected daily peak requirement.

■ The capacity to contribute via appropriate investments plans to the development and security of the gas system or supply diversification.

The MAP will be responsible for emergency planning and safety conditions of the Italian system. The Ministry will supervise long-term planning and may give specific instructions to safeguard continuity and security of supply and the functioning of the storage system. In case of an energy crisis or serious risks, the Ministry will decree necessary short-term measures, which are to be kept to the strict minimum.

- **Access to transportation, storage and LNG terminals**

Access to all facilities is regulated, except for newly-built LNG terminals. The energy regulator (Autorità per l'energia elettrica e il gas) has published the criteria for setting transmission, dispatching and local distribution tariffs. The energy regulator has also published tariffs for access to storage and LNG terminals.

A decree was published in April 2001 to regulate third-party access to depleting gas fields, for the purpose of converting them into storage facilities.

According to Order n. 91/02 of 16 May 2002, access to newly-built LNG facilities is negotiated. The sponsor of a newly-built LNG terminal has priority of access to the terminal, with the following constraints: up to 20 years of duration; up to 80% of financed new LNG regasification capacity; eligibility to the priority of access is limited to 8.3 bcm/year of new capacity and to up to 25 bcm/year of total regasification capacity. A negotiated price applies for such a priority access. Use-it or lose-it will also apply for the priority of access, on a yearly basis. For other customers, the remaining capacity (20%) and any non-assigned capacity on a priority basis will be made available at tariffs set by the Regulatory Authority.

JAPAN

- Gas represents 12% of Japan's energy mix. The country relies on LNG imports for 97% of its gas supplies. About two thirds of LNG imports are used for power generation and most of the rest to make city gas.

- Seasonal fluctuations in gas demand are less pronounced than in other regions.

- Japan has developed extensive flexibility tools to cope with its unique situation. LNG, imported under long-term contracts, comes from eight countries and ten LNG plants.

- LNG tanks at regasification terminals provide ample storage capacity.

- Forty per cent of Japan's gas-fired power plants are multi-fired and can switch to other fuels.

GAS DEMAND

- **Share of gas in TPES:** 12% (2000)

Gas consumption reached 78 bcm in 2000. Almost all gas consumed in Japan is imported as LNG. Japan initiated LNG trade in the Asia-Pacific region in 1969 with its first imports coming from Alaska. The country has since become the world's largest LNG importer.

About two-thirds of LNG imports are used for power generation. Although Japan is the seventh biggest gas consumer in the world, it has a very limited gas

Figure 54: Japanese gas consumption by sector in 2000

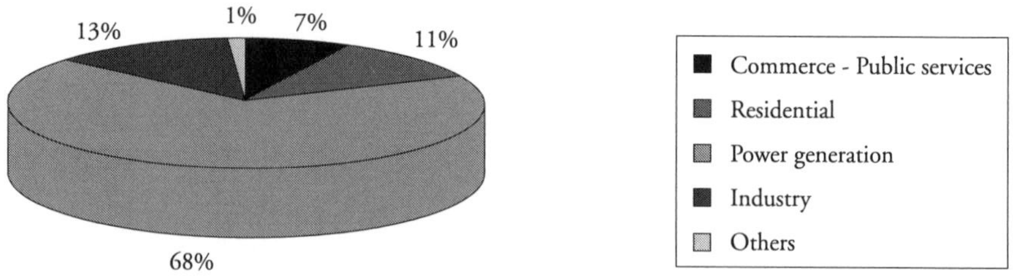

13%   1%   7%   11%

68%

- Commerce - Public services
- Residential
- Power generation
- Industry
- Others

transmission system. Deregulation will help to open up new markets for city gas (residential/commercial sectors). This could have an effect on the seasonality of gas demand.

■ Seasonality

The ratio between gas sales in the peak and the lowest month of the year was 1.4 to 1 in 2000. There are two peaks in Japan: one in the winter for heating purposes and one in summer for air-conditioning.

Figure 55: Japanese seasonality in gas demand

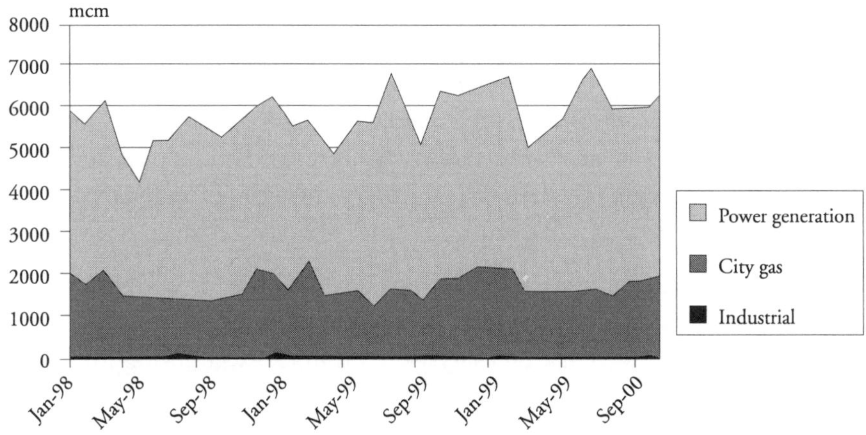

Source: www.eneken.ieej.or.jp

■ Share of gas in the power generation sector: 22% (2000)

Table 35: Multi-fired electricity generating capacity in Japan at 31 December 2000 (GW)

Solid/Liquid	3.01
Solids/Gas	0
Liquids/Gas	23.36
Liquids/Solids/Gas	0
Total multi-fired	26.37
Total capacity*	166.65

* from combustible fuels.

16% of total electricity-generating capacity by fuel is multi-fired, with plants running on natural gas playing a leading role.

Maximum electricity-generating capacity from gas was 58.62 GW at the end of 2000. Gas used for electricity generation is projected to increase from around 59 GW to about 67 GW by 2010, though its proportion of total capacity will remain stable. Coal and nuclear power will expand further than gas. Since nuclear and coal will be used for base load, gas may move slightly up the merit order. At present gas-fired plants operate for about 4,000 to 5,000 hours a year. This load factor is expected to decline slightly. As dependence on oil from power generation is reduced, gas may increasingly need to be used to meet peak demand, making consumption less regular and predictable.

- **Interruptible contracts and fuel switching**

Forty per cent of gas-fired power generating capacity is dual-fired, with crude or fuel oil as the main alternative fuel. Fuel switching would pose few logistical problems, as the sites are all coastal and have storage and handling capacity. This flexibility will decline somewhat in the future, as new gas-fired generation will be mainly single-fired CCGT plants.

For city-gas contracts there is less flexibility. There are no interruptible contracts as such, though contracts with industrial consumers contain *force majeure* clauses. Only about 20% of larger city-gas consumers, accounting for a small proportion of total demand, have dual-firing and that proportion is declining.

GAS SUPPLY

- **Reserves:** 40 bcm   **R/P:** 16 years
- **Gas supply structure:** indigenous production 3%; imports 97% of which:
    - Indonesia 32%
    - Malaysia 20%
    - Australia 13%
    - Qatar 11%
    - Brunei 10%
    - UAE 8%
    - USA 2%
    - Oman < 1%

Japan has diversified its LNG imports, which now come from eight countries and ten LNG plants. Oman started to deliver in December 2000.

- **LNG share in gas supply:** 97%

- **Supply swing**

Figure 56: Japanese monthly gas supply

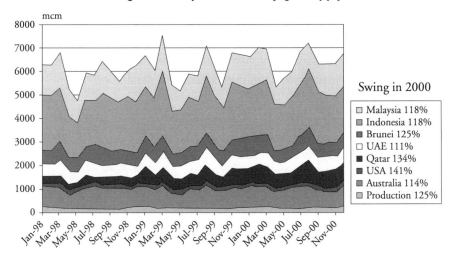

LNG supplies offer little swing (113% in average), except for LNG received from Qatar, with swing of 134% in 2000. This is due to the build-up of Qatar plants and available extra capacity.

STORAGE

There is no underground gas storage in Japan. However, the country has many LNG aboveground tanks at its LNG terminals.

- **LNG import terminals**

Japan has 22 regasification terminals, for a total capacity of 603.8 mcm/day, or about 200 bcm/year, which gives the country a huge spare capacity. The storage capacity is 12.18 mcm of LNG (equivalent to 7.3 bcm of gas), which is huge in comparison to other LNG importing countries, but this is the only means to store gas in Japan.

Many terminals serve specific power plants but some are shared between electricity and city-gas companies.

Table 36: Japanese LNG import terminals

Terminals	Storage tanks (1,000 cm of LNG)	Nominal capacity (mcm/day)	Start-up date
Negishi	1,250	50.8	1969
Senboku I	180	8.4	1972
Sodegaura	2,660	103.6	1973
Senboku II	1,405	43.8	1977
Tobata	480	24	1977
Chita I	300	27	1977
Himeji I	520	31.6	1979
Chita II	640	42.9	1983
Higashi-Niigata	720	31.4	1983
Himeji II	560	14.8	1984
Higashi-Ohgishima	540	62.9	1984
Futtsu	610	69.3	1985
Yokkaichi (Kawagoe)	320	29.2	1987
Yanai	480	7.5	1990
Shin-Oïta	320	17.2	1990
Yokkaichi	160	2.4	1991
Fukuoka	70	1.7	1993
Hatsukaichi	85	1.3	1996
Sodeshi	82.9	2.3	1996
Kagoshima	36	0.5	1996
Shin-Minato	80	1.1	1997
Kawagoe	480	19.4	1997
Ohgishima I	200	10.7	1998
Total	**12,178**	**603.8**	

■ **Stock changes**

The graph below shows that gas consumption is relatively flat throughout the year (load of 112%). Fluctuations in gas demand are covered by import flexibility.

Figure 57: Japanese load balancing

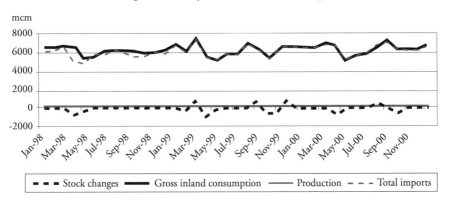

mcm

Legend: Stock changes — Gross inland consumption — Production — Total imports

REGULATORY FRAMEWORK

- **Security of supply**

Japan's gas market is different from those in the other IEA regions. Japan is almost completely dependent on imported LNG, most of which is used for power generation.

The special structure of the market determines Japan's approach to energy security. Underground storage, interruptible contracts, pipeline links and other approaches that are common in other regional markets play no significant role in Japan. Reliance is placed on long-term contracts with several stable suppliers, on modular supply and delivery systems that limit dependence on any single installation, and on fuel substitution and sharing via the electricity-generation system.

These arrangements have served Japan well and no serious security problems have been encountered, even when the Arun plant in Indonesia ceased to deliver from March to July 2001.

The Japanese approach to security of supply is very comprehensive. It seeks to:

1. conserve resources and alleviate energy loads through energy conservation;

2. obtain the best mix of energy supply, through diversification of supply sources and the use of more-environment-friendly fuels;

3. decentralise supply sources and promote producer-consumer dialogue;

4. promote security of the environment;

5. take measures for emergency preparedness, especially stockpiling;

6. take measures to curb price volatility, by establishing markets that minimise the perverse effects of deregulation.

In the gas sector, Japan has a series of measures providing insurance against supply interruptions:

■ *Supply diversity:* Eight countries supply LNG to Japan. Individual Japanese companies generally have more than one supplier. Osaka Gas, for example has six suppliers, under nine separate contracts.

■ *Long-term contracts:* Suppliers and customers are interdependent and have a common interest in security of supply. They are linked by long-term contracts that have proved a stable basis for managing the business in the past.

■ *Modular supply systems:* Production and liquefaction plants include a number of separate units; several tankers are involved in each contract; most importing companies have more than one terminal; terminals have more than one jetty.

■ *Supply flexibility.* Most supply contracts have from 5% to 10% flexibility either written into the contract or on a "best endeavours" basis.

■ *Gas supply sharing:* Although there are few pipeline connections, a number of terminals are shared between gas and electricity companies. Furthermore, there is a high degree of standardisation of shipping capacity: extra supply available from a particular source can usually be transferred to another company that might be facing difficulties.

■ *Electricity exchanges:* Japan has two frequency zones - 50 Hz and 60 Hz. Electricity interconnections exist both between companies within the zones and, to a limited extent, between the zones. These interconnections are being expanded.

■ *Fuel-switching:* Forty per cent of gas-fired power generating capacity is dual-fired, with crude or fuel oil as the main alternative fuel.

■ *SNG manufacture:* There is considerable capacity for manufacturing synthesised natural gas (SNG) from naphtha; capacity is around 1.4 million tons annually for city-gas companies as a whole.

■ *Storage:* Although Japan has no underground storage, it has large above ground capacity designed to cope with fluctuations in supply. Total storage, at 7.3 bcm, amounts to 34 days of average consumption.

■ Access to the grid

Japan's Ministry of Economy, Trade and Industry (METI) plans to force gas suppliers to allow third-party access to their gas systems by March 2004. The move will apply to both the retail and wholesale gas markets.

Initial steps towards competition in the Japanese electricity and city-gas markets have been gradually taken since 1994. In 1999, the requirements for eligible customers for liberalisation were decreased from purchases of 2 mcm/year to over 1 mcm/year. The deregulated segment of the city gas market is 30% of the total, and of the utilities market 20%. The deregulation of the electricity sector, started in 2000, is also affecting the gas market. City-gas and electricity utilities can now sell in one another's territory.

KOREA

- Korea imports all its gas from LNG suppliers. The country is the world's second largest LNG importer with 19 bcm imported in 2000. Korea does not produce gas.

- The temperature-sensitive market represents half of gas sales. Demand seasonality is high.

- Seasonal variations in sales are mainly met by import flexibility and spot LNG cargoes imported in the winter period.

- There is no underground gas storage on the territory.

- LNG facilities have storage tanks and provide some flexibility.

GAS DEMAND

- **Share of gas in TPES:** 9% (2000)

Korean gas consumption reached 19 bcm in 2000. Natural gas is principally consumed by the power and residential sectors. The residential sector represents 39% of total consumption.

Until the economic crisis at the end of 1997, Korea's LNG use grew by an average 25% per year in the 1990s. Imports decreased in 1998, but started to rise again in 1999, by 23% and in 2000 by 12%.

Figure 58: Korean gas consumption by sector in 2000

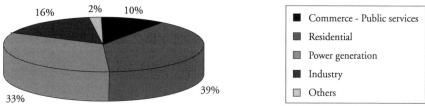

16% 2% 10%

33%

39%

- ■ Commerce - Public services
- ▨ Residential
- ▧ Power generation
- ■ Industry
- ☐ Others

■ Seasonality

The ratio between gas sales in the peak and the lowest month of the year was 2.8 to 1 in 2000. As can be seen from the graph below, consumption in the residential sector is highly temperature-dependent, with gas sales in December 2000 about ten times higher than in August 2000. Gas consumption in the commercial sector is also highly seasonal.

Gas consumption in the electricity sector peaks in the summer months and thus helps smooth out seasonal variations in gas demand.

Figure 59: Korean gas demand seasonality[64]

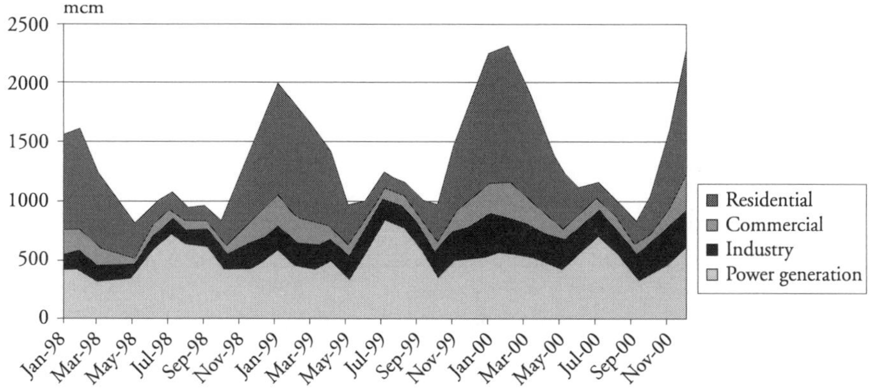

Source: Korea Energy Economics Institute.

■ Share of gas in the power generation sector: 10% (2000)

**Table 37: Multi-fired electricity generating capacity in Korea
at 31 December 2000 (GW)**

Solid/Liquid	1.61
Solids/Gas	0
Liquids/Gas	6.97
Liquids/Solids/Gas	0
Total multi-fired	8.58
Total capacity*	36.82

* from combustible fuels.

---

64 Includes LPG (3% of total consumption in 2000).

23% of total electricity generating capacity by fuel is multi-fired, with plants running on natural gas playing the largest role.

GAS SUPPLY

- **Gas supply structure:** indigenous production zero; imports 100%, of which:
    - Indonesia 42%
    - Qatar 22%
    - Malaysia 17%
    - Oman 11%
    - Brunei 5%
    - United Arab Emirates 2%
    - Others - spot 1%

LNG is imported under long-term contracts from Indonesia, Malaysia, Oman, Qatar and Brunei, and also under short-term contracts from Abu Dhabi.

- **LNG share in gas supply:** 100%

- **Supply swing**

LNG supplies offer a great deal of flexibility as the deliveries under long-term contracts are complemented by spot cargoes. Additional short-term LNG trade to meet peak demand in the winter months of 2001-2002 required 16 cargoes.

Figure 60: Korean monthly LNG supplies

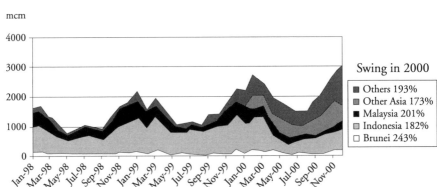

STORAGE

Korea does not have underground gas storage.

■ LNG import terminals

Korea Gas Corporation (Kogas) has two LNG regasification terminals, with a storage capacity of 2 mcm of LNG (1.2 bcm).

Table 38: Korean LNG import terminals

Terminals	Storage tanks (1,000 cm of LNG)	Nominal capacity (mcm/day)	Start-up date
Pyeong Taek	1,000	60	1986
Incheon	1,000	77	1996

Construction of a third Kogas LNG terminal began in 1999, at Tongyong, in the south of the country, west of Pusan. It is due to come onstream at the end of 2002 with a 3-mt/year capacity in the first phase. Posco of Korea has plans for an LNG receiving terminal at Kwang Yang to supply gas to its power stations in Kwang Yang and Pohang.

■ Stock changes

Demand seasonality is mainly met by import flexibility. The swing in imports is 147%. LNG regasification tanks also offer some flexibility.

Figure 61: Korean load balancing

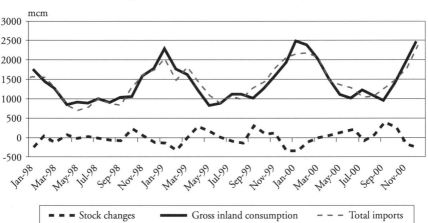

REGULATORY FRAMEWORK

■ **Security of supply**

One of the major objectives of the Korean energy policy is to maintain a diverse and stable energy supply. LNG was introduced in 1986 as one of the pillars of Korea's energy-diversification policy. LNG is imported under seven long-term contracts with five major suppliers.

Security of supply was tested in 2001 when Exxon-Mobil's Arun LNG complex in North Aceh, Indonesia, was forced to cease production from March to September. The shut down of the Indonesian gas fields had a direct short-term impact in Korea, as Arun had contracted to provide Korea 3.3 million tonnes of LNG per annum. During the shutdown, additional supply was made available by Bontang in Indonesia and by Malaysia and Australia. This was made possible by the existing spare capacity at these plants. It demonstrated the excellent co-operation that exists among LNG players of the region.

■ **Gas market reforms and access to the LNG terminals/grid**

The government will liberalise and privatise the LNG import, transmission and wholesale businesses, which are now a monopoly of state-owned Kogas.

The basic steps of the restructuring plan include:

■ introducing gas-to-gas competition by unbundling of importing and sales activities from operation of terminals and the transmission network;

■ instituting an open access regime for receiving terminals and the transmission network;

■ introducing competition in the retail sector through competition in facility investment.

The plan includes dividing Kogas into four companies: a holding company for infrastructure (terminals, pipelines and trunk line) and three trading companies. Two of these import and wholesale subsidiaries would be sold off by the end of 2002. The holding company and the third trading company would also be sold off, but not immediately. LNG purchasing agreements will be grouped so as to ensure fair and transparent competition between the three import and wholesale companies. The three companies will each undertake three businesses: LNG imports, wholesaling and retailing.

LUXEMBOURG

- Luxembourg has a small gas market.
- Gas is mostly used by the industrial sector.
- Imports from Norway provide the required seasonal flexibility.
- Balancing services are in place.

GAS DEMAND

- **Share of gas in TPES:** 18% (2000)

Luxembourg's gas consumption was 0.76 bcm in 2000. Most of the gas was consumed by the industrial sector.

Figure 62: Luxembourg gas consumption by sector in 2000

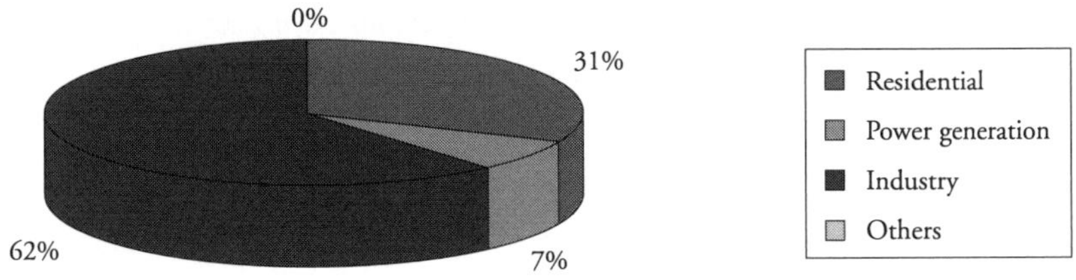

0%

31%

62%

7%

- Residential
- Power generation
- Industry
- Others

- **Seasonality**

The ratio between gas sales in the peak and the lowest month of the year was about 2 to 1 in 2000.

- **Share of gas in the power generation sector:** 53% (2000)

- **Interruptibles and fuel switching**

Gas power plants are single-fired.

GAS SUPPLY

■ **Gas supply structure:** indigenous production zero; imports 100%, of which:
  - France 6%
  - Germany 2%
  - Norway 92%

■ **Supply swing**

Figure 63: Luxembourg monthly gas supply

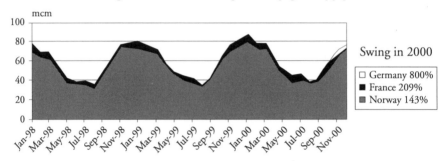

Norwegian imports which provide the bulk of gas supplies to Luxembourg, provide the required flexibility, with a swing of 143%.

STORAGE

There is no gas storage in Luxembourg.

■ **Stock changes**

Figure 64: Luxembourg load balancing

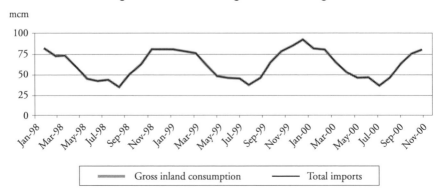

Demand seasonality is not very high in comparison with other European countries and is met entirely by import flexibility.

REGULATORY FRAMEWORK

■ **Access to the grid and storage**

Grid access conditions are regulated and tariffs are published on a temporary basis until the regulator approves them. Balancing services are in place and risk management is possible, but the large buyers have not used it so far.

NETHERLANDS

- Natural gas accounted for 46% of the Netherlands' total primary energy supply in 2000, a higher proportion than in most other OECD countries.

- Seasonal variations in gas demand are less pronounced in the Netherlands than in other European countries. The residential and commercial sectors represent only 25% of total sales.

- Supplies to electricity generators can be interrupted when temperatures drop below 0°C.

- Thanks to the giant Groningen gas field and its proximity to European markets, the Netherlands is the traditional European swing-gas supplier. The high swing offered by Groningen enables the Dutch gas industry to meet gas-demand fluctuations.

- Until recently the Netherlands had only one LNG peak-shaving facility. Currently, Gasunie operates three underground storage sites and BP plans to develop a new site.

- Gasunie offers access to "virtual storage" and plans to offer storage services in the future.

GAS DEMAND

- **Share of gas in TPES:** 46% (2000)

Gas consumption reached 48.9 bcm in 2000. The residential sector's share in overall gas consumption remains very stable, at 25%, whereas power generation's share is rising.

Figure 65: Dutch gas consumption by sector in 2000

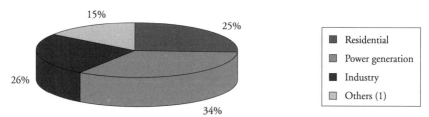

(1) Agriculture accounts for 43% and the energy sector for 10% of the "others" category.

■ **Seasonality of gas demand**

The ratio between gas sales in the peak and low month of the year was 2.3 to 1 in 2000.

■ **Share of gas in the power generation sector:** 58% (2000)

Gas has a very high level of penetration in the power sector.

Table 39: Multi-fired electricity generating capacity in the Netherlands at 31 December 2000 (GW)

Solid/Liquid	0.60
Solids/Gas	3.57
Liquids/Gas	3.91
Liquids/Solids/Gas	0
Total multi-fired	8.08
Total capacity*	20.07

* from combustible fuels.

40% of total electricity generating capacity by fuel is multi-fired, with plants running on natural gas playing the biggest role.

■ **Interruptible customers and fuel switching**

Supplies to power stations can be interrupted when temperatures fall below zero degree Celsius. They are routinely interrupted when temperatures drop below minus 5°C.

GAS SUPPLY

■ **Gas reserves:** 1,680 bcm   **R/P:** 25 years

■ **Gas supply structure:** indigenous production 81%; imports 19%, of which:
  - UK 11%
  - Norway 7%
  - Germany 1%

■ Supply swing

Figure 66: Dutch monthly gas supply

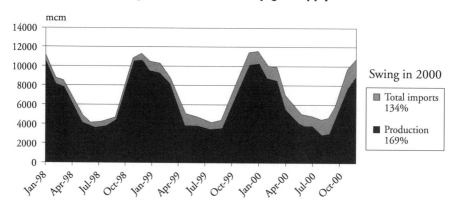

The Netherlands has the second-largest gas reserves in the IEA-European countries, much of them in the Groningen field. In 2001, Groningen supplied just over 29% of the total volume of gas purchased by Gasunie.

Production from Groningen rose exponentially between 1963 and 1973. In 1974, the Dutch government introduced the "small gas fields" policy, designed to discover, develop and operate smaller fields in order to improve the management of its long-term natural gas resources. Under this policy, Gasunie is obliged to buy gas produced at small fields, when available, rather than from Groningen. The policy was facilitated by the fact that Groningen has unique flexibility, with a deliverability ranging between 0 and 500 mcm per day. This is sufficient to cope with Dutch and European customers' demand seasonality. In effect, Groningen acts as a swing field.

Dutch production will eventually decrease as will its contribution to meeting flexibility in demand. Imports from Norway and UK are increasing (up 140% in three years) but these offer very little flexibility. With the opening of the Interconnector, imports from the UK began to augment, bringing intensified competition to the Dutch market.

STORAGE

For a long time, Holland had no underground gas storage at all, only a peak-shaving facility used during periods of very cold weather. The Groningen field could, and still can, balance seasonal variations in consumption and play the role of a swing producer.

Table 40: Underground gas storage in the Netherlands

Name	Type	Operator/ Number	Working capacity (mcm)	Peak output (mcm/day)
Maasvlakte	LNG peak shaving	Gasunie	78	31
Norg (operated by NAM)	Depleted gas field	Gasunie	1,100	54
Grijpskerk (NAM)	Depleted gas field	Gasunie	800	54
Alkmaar (Amoco)	Depleted gas field	Gasunie	500	36
Total		**4**	**2,478**	**175**

Faced with the prospect of the depletion of the Groningen field, Amoco, NAM/Gasunie and BP converted depleted gas fields into storage reservoirs. These facilities provide new flexibility. They also permit delaying the installation of extensive compression facilities on the Groningen field.

Gasunie contracted all capacity at the country's three storage facilities on a long-term basis. There is open access to Gasunie's storage, but the price of access is not published. There is no secondary storage market, and the Dutch energy regulator is now reviewing the situation.

There is also an independent gas storage project, developed by BP and its partners in the P15-P18 gas field complex, some 40 kilometers offshore Rotterdam. The complex can currently ship 15 mcm/day of gas via a dedicated 26-inch pipeline to shore at Maasvlakte, where gas can be sold to industrial or oil-refining clients, or shipped onward via Gasunie's open-access transmission network to Rotterdam.

■ **Stock changes**

Unlike other European countries, the Netherlands did not develop storage to balance gas supply and demand. Seasonal variations in gas demand are met by swing production. Storage was developed to allow Gasunie to continue buying gas from small fields with high load factors in priority over Groningen. Gasunie can fulfil this obligation by adjusting purchases from Groningen and using storage.

Figure 67: Dutch load balancing

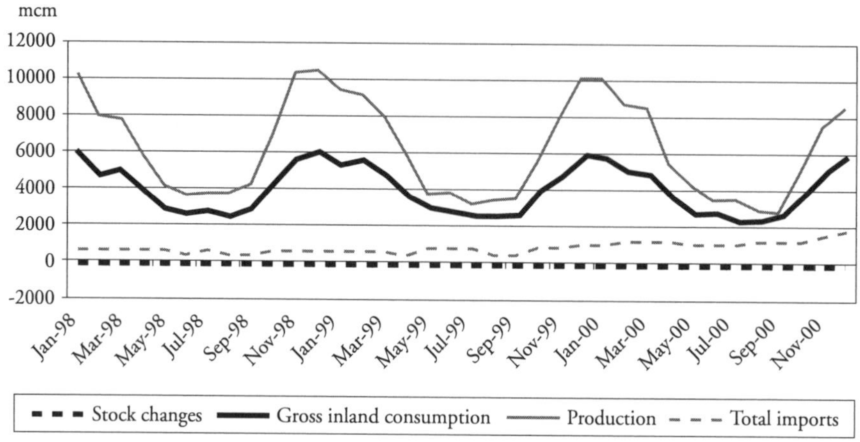

Legend: Stock changes — Gross inland consumption — Production — Total imports

REGULATORY FRAMEWORK

■ **Security of supply**

Security of supply is left to the markets. Suppliers to small households must have permits based on guaranteed supply to these markets.

Gasunie has designed its pipeline network so that it can cover demand on days when the average temperature drops below minus 15°C (which happens once every 50 years, according to national statistics).

■ **Access to transportation and storage**

The Netherlands opted for a negotiated third-party access for transmission. Access, however, is subject to regulatory control. Indicative tariffs and terms and conditions for transport and necessary ancillary services are published. Services associated with transportation include load-factor conversion, providing flexibility and services relating to gas quality and pressure.

Storage companies with a dominant position (NAM, BP Amoco and Gasunie) are required to give new parties access to their gas-storage installations.

■ **Gasunie services**

Gasunie Trade & Supply (GTS) provides the following services: transportation, quality conversion and hourly flexibility.

■ *Transport services.* GTS provides transportation services for non-Gasunie gas from the point of delivery to the point of redelivery. The company uses its 12,000-km pipeline system, which covers all of the Netherlands. These services are provided in accordance with the following principles:

- first come, first served;

- declaration of contract;

- hourly balancing (hour in = hour out).

■ *Quality conversion.* Gasunie can adjust the quality of the gas to the requirement of its customers.

■ *Hourly flexibility.* GTS provides hourly flexibility. It offsets short-term imbalances by providing *virtual storage* at the consumer location. Hourly flexibility is a virtual gas storage tank that is made available to the customer. This tank has a volume and a given send-out capacity. Hourly flexibility entitles the customer to have capacity available for a certain period in addition to capacity supplied by another gas supplier.

NEW ZEALAND

- Gas consumption, which is entirely covered by indigenous production, reached 6.1 bcm in 2000. Gas sales to the industrial sector, especially petrochemicals, account for 52% of total sales. Sales to the power sector account for 41%.

- Seasonal variations in sales are low and are entirely met by flexibility in production.

- New Zealand has no gas storage facility.

GAS DEMAND

- **Share of gas in TPES:** 27% (2000)

New Zealand's gas consumption reached 6.07 bcm in 2000. There are three major consumer groups: the petrochemicals industry, electricity generation and direct users served by gas utilities. In 2000, 42% of New Zealand's natural gas was used in the production of methanol at the Montunui and Waitara plants, both owned by Methanex New Zealand Limited. Natural gas is also used as a fuel and feedstock in the manufacture of ammonia and urea fertilisers. A further 41% of the gas was used for electricity generation. The remaining 17% was delivered through a high-pressure pipeline system directly to major users, to gas utilities for distribution to other industrial users and to the commercial and industrial sectors.

Figure 68: New Zealand gas consumption by sector in 2000

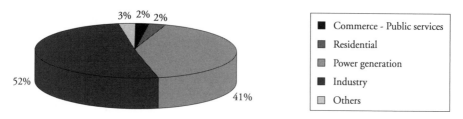

3% 2% 2%
52%
41%

- Commerce - Public services
- Residential
- Power generation
- Industry
- Others

Only the North Island has a gas distribution system.

Due to depletion of Maui field, gas consumption is expected to decrease. According to the long-term outlook developed by the Ministry of Economic Development, even in a high "gas discovery scenario", the existing petrochemicals plants will have to close in the future.

■ **Seasonality**

The ratio between gas sales in the peak and the lowest month of the year was 1.2 to 1 in 2000.

■ **Share of gas in the power generation sector:** 24% (2000)

Table 41: Multi-fired electricity generating capacity in New Zealand at 31 December 2000 (GW)

Solid/Liquid	0
Solids/Gas	1
Liquids/Gas	0
Liquids/Solids/Gas	0
Total multi-fired	1
Total capacity*	2.81

* from combustible fuels.

36% of total electricity-generating capacity by fuel is multi-fired.

GAS SUPPLY

■ **Reserves:** 62 bcm   **R/P:** 11 years

■ **Gas supply structure:** indigenous production 100%; imports zero

All New Zealand's gas is produced in the Taranaki region, mainly from the Maui field. The Maui field is owned by Energy Exploration NZ Ltd, Shell Petroleum Mining and Todd Energy. The Kapuni field is owned by Shell Petroleum Mining and Todd Energy. In 2000, the Maui and Kapuni fields produced 91.1% of all the country's gas production. The Maui field is depleting rapidly.

■ **Supply swing**

Gas production offers little swing (110% in 2000).

Figure 69: New Zealand monthly gas production

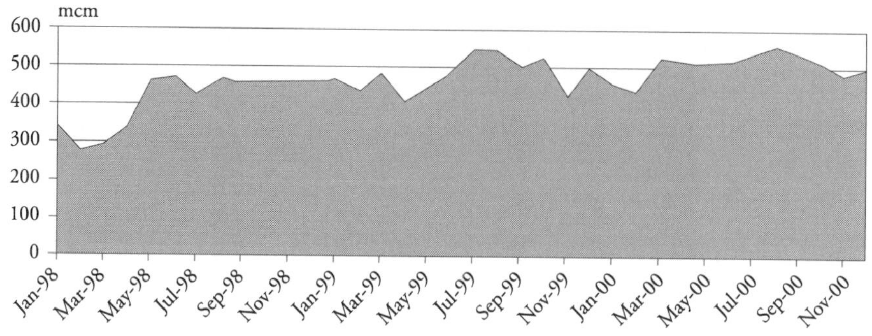

mcm
600
500
400
300
200
100
0
Jan-98
Mar-98
May-98
Jul-98
Sep-98
Nov-98
Jan-99
Mar-99
May-99
Jul-99
Sep-99
Nov-99
Jan-00
Mar-00
May-00
Jul-00
Sep-00
Nov-00

STORAGE

There is no gas storage in New Zealand.

■ Stock changes

Demand seasonality is low and entirely met by swing in indigenous production.

Figure 70: New Zealand's load balancing

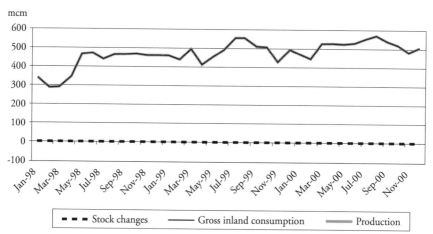

mcm
600
500
400
300
200
100
0
-100
Jan-98
Mar-98
May-98
Jul-98
Sep-98
Nov-98
Jan-99
Mar-99
May-99
Jul-99
Sep-99
Nov-99
Jan-00
Mar-00
May-00
Jul-00
Sep-00
Nov-00
Stock changes
Gross inland consumption
Production

REGULATORY FRAMEWORK

■ **Security of supply**

As the country is self-sufficient in gas, security of supply means mainly coping with technical interruptions in gas supply. This is provided by the integrated control system for high-pressure transmission lines, line-pack and interruptible supply to large customers (mainly electricity generators). An industry-agreed rationing programme is also available if necessary.

In the longer term, the depletion of the Maui field will create new challenges for security of supply.

■ **Access to the grid**

A Gas Pipeline Access Code came into effect in early 1999. It is a voluntary, self-regulatory agreement between pipeline owners and users. It covers all pipelines except gas-gathering pipelines. The code provides for neutral, non-discriminatory access to available capacity. It is voluntary, has no sanctions, does not affect pricing and relies on information disclosure regulations.

NORWAY

- Norway is a major producer and exporter of natural gas.

- Norway offers flexibility to its European customers under its long-term contracts.

- The country has a little developed onshore gas market.

- Large amounts of gas are reinjected in oil fields.

- Regulatory reforms of the gas sector are taking place.

GAS DEMAND

- **Share of gas in TPES:** 14% (2000)

Norwegian gas consumption reached 4.2 bcm in 2000. Norway has not yet developed its onshore gas market. Its current gas consumption consists mainly of gas used for the compression and treatment of gas. Large amounts of gas (35 bcm in 2000) are reinjected in oil fields for secondary recovery.

Since the mid-1980s, several alternatives for gas-fired power generation in Norway have been commercially evaluated, but so far, none of these projects have been realised. In 1994, Statkraft, Statoil and Norsk Hydro set up a joint company, Naturkraft. The objective of Naturkraft is to use natural gas from the Norwegian Continental Shelf for generation of electric power for the Nordic market.

The Norwegian Government is preparing a white paper for the Parliament about onshore domestic use of natural gas. Opportunities exist for direct use of natural gas in industry, transport and for stationary energy purposes. Natural gas will have positive effect on the environment if it is used to substitute other, more pollutive fossil fuels. Natural gas could also allow the substitution of gas for electricity in some applications and, in some instances, gas transmission might effectively replace the need for expanding the electricity grid.

GAS SUPPLY

■ **Reserves:** 4,017 bcm　　**R/P:** 76 years

■ **Gas supply structure:** indigenous production 100%; imports zero

Marketed gas production reached 52.8 bcm in 2000.

■ **Supply swing**

Figure 71: Norwegian monthly gas production

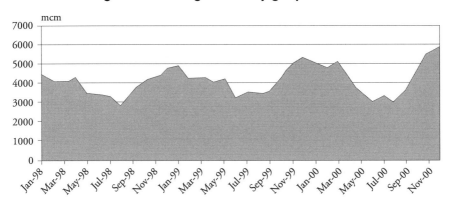

Norway offers flexible supplies under its long-term contracts, as seen in the above graph. The swing factor was 124% in 1993 and reached 132% in 2000. Troll - which accounts for approximately half of Norwegian sales - and Sleipner offer a great deal of flexibility.

All recent long-term supply contracts from small Norwegian fields to buyers in continental Europe have been so-called Troll Gas Sales Agreements that ensure a customer's allotted supply is met using Troll gas when a smaller field is shut in.

The flexibility of Norwegian gas supply is well illustrated by the fact that 2001 saw 791 field or pipeline shut-downs, and still the regularity of supply was 99.8%.

■ **Gas exports:** Norway exported 48.5 bcm in 2000:

- Germany 38.2%

- France 24.7%

- Belgium 11.3%

- Netherlands 10.5%

- Spain 5.1%

- UK 4.5%

- Czech Republic 3.9%

- Austria 1.7%

- Poland 0.1%

STORAGE

Norway has no gas storage on its territory. But Norwegian gas sellers can accommodate customers by drawing on storage capacity on the Continent. This capacity is partly owned by Norwegian companies (like Etzel gas storage in northern Germany) and partly rented (like GDF storage in France).

REGULATORY FRAMEWORK

■ **Security of supply**

The Norwegian integrated resource management policy is designed to ensure efficiency in production and transportation of natural gas. The Norwegian Petroleum Directorate is charged with overseeing the technical aspects of security of supply on the Norwegian Continental Shelf.

A comprehensive policy assessment of state ownership in the offshore oil and gas sectors took place in 2001. Statoil was partially privatised while the State Direct Financial Interest was restructured. The Gas Negotiating Committee (GFU), which has co-ordinated the country's exports since its creation in 1993, terminated its activities as from 1 June 2001, and was formally abolished on 1 January 2002. Gas producers on the Norwegian Shelf now individually market their own gas. Changes in policy on oil and gas development and gas marketing, are driven by the ongoing changes in the European gas markets and the maturation of the Norwegian Continental Shelf.

Norway has developed an extensive transportation infrastructure, including six major pipelines carrying gas from the Norwegian Continental Shelf to

Continental Europe and the United Kingdom. Major changes are taking place in the gas transportation system. The state owned company Gassco has taken over the operatorship of the gas transportation system from 1 January 2002. Gassco is a neutral operator for the integrated transportation system and is responsible for operating the major gas transport pipelines in an efficient and secure way.

The transmission system offers a great deal of flexibility, as was demonstrated in April 2000, when the Zeepipe gas trunkline from Norway to Zeebrugge in Belgium was shut down for two weeks following a gas leak upstream. The pipeline carries about one fifth of all Norwegian gas exports. Customers for Zeepipe gas were offered supplies from alternative landfalls in Germany and France.

■ **Access to upstream pipelines**

There is negotiated access to the Norwegian upstream gas transportation systems, with the government having a reserve power under the Petroleum Act to determine prices in the event of a dispute about terms. All gas transportation agreements must be approved by the Government.

Tariffs in the upstream gas transportation pipelines are regulated by the Ministry of Petroleum and Energy under the existing law. Tariffs for the gas export infrastructure conveying gas to Europe are cost-based, with a 7% real rate of return before tax as the guideline for determining tariff levels.

A process of unitisation of ownership in pipelines and a government evaluation of the access regime are taking place. The Ministry of Petroleum and Energy aims at issuing regulations for a new access regime for upstream pipelines before end 2002.

PORTUGAL

- Portugal is an emerging gas market. It is still very small.

- Gas is principally used in power plants and industry. Seasonal variations of gas demand are limited.

- Gas is imported from Algeria by pipeline under long-term contracts.

- Flexibility in Algerian imports covers the seasonal fluctuations in gas demand.

- There are no storage facilities yet. An LNG terminal is under construction in Sines.

- As an emergent market, Portugal applied for a derogation to the EU Gas Directive until 2007.

GAS DEMAND

- **Share of gas in TPES:** 8% (2000)

Portugal's gas consumption reached 2.3 bcm in 2000, most of it for power generators and industrial customers.

According to forecasts by the Ministry of Economy, natural gas supply is expected to increase at an annual rate of 19% per year until 2010 when it will reach 5.7 Mtoe, 22.7% of total energy supply. Gas use in power generation is expected to be multiplied by more than ten between 1998 and 2010, reaching 3.9 Mtoe (more than 68% of total gas supply) in 2010.

Up to 2010, gas consumption in industry and in the commercial and residential sectors will be multiplied by approximately five to six. Industry will continue to be the second-largest gas-consuming sector with 1.2 Mtoe in 2010. Competition is expected to become sharper in these sectors once natural gas has replaced propane and butane in the distribution areas where natural gas has a competitive price advantage.

Figure 72: Portuguese gas consumption by sector in 2000

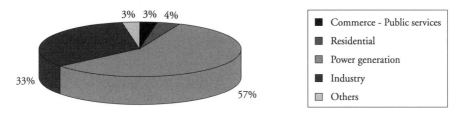

3% 3% 4%

33%

57%

- ■ Commerce - Public services
- ■ Residential
- ▨ Power generation
- ■ Industry
- □ Others

■ **Seasonality**

The ratio between gas sales in the peak and the lowest month of the year was 2 to 1 in 2000.

■ **Share of gas in the power generation sector:** 16% (2000)

Table 42: Multi-fired electricity generating capacity in Portugal at 31 December 2000 (GW)

Solid/Liquid	0.21
Solids/Gas	-
Liquids/Gas	0.71
Liquids/Solids/Gas	-
Total multi-fired	0.92
Total capacity*	6.27

* from combustible fuels.

15% of total electricity-generating capacity by fuel is multi-fired, with plants running on natural gas playing a significant role.

GAS SUPPLY

■ **Gas supply structure:** indigenous production zero; imports 100%, of which:

- Algeria 87%

- Nigeria 10%

- Malaysia 3%

Algerian gas is imported by pipe via Spain. LNG imports are received at Spanish terminals, pending the opening of the Sines regasification plant.

■ **Supply swing**

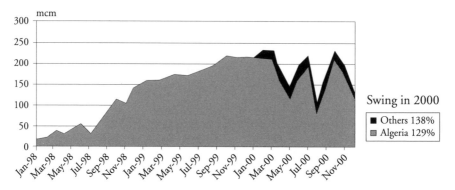

Figure 73: Portuguese monthly gas supplies

Others include Malaysian and Nigerian LNG received at Spanish plants.

STORAGE

Currently, there are no gas storage facilities in Portugal.

■ **Stock changes**

Figure 74: Portuguese load balancing

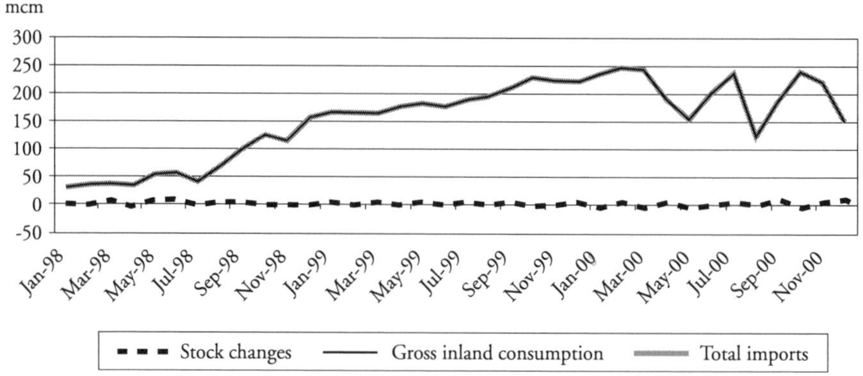

As indicated in the graph, there is little seasonality in demand. What there is is met by flexibility in imports.

REGULATORY FRAMEWORK

■   **Access to transportation and storage**

Portugal is an emergent market and has applied for a derogation from the EU Gas Directive, which Portugal will not implement until 2007.

SPAIN

- Spain's gas consumption reached 17 bcm in 2000. Sales to the temperature-sensitive market are limited, amounting to 17% in 2000.

- Seasonal fluctuations in gas demand are limited. The ratio between gas sales in the peak and the lowest month of the year was about 2 to 1 in 2000.

- Some industrial customers, as well as power-generation plants, have interruptible contracts.

- Flexibility in gas supply is limited. Indigenous production is negligible. Storage is limited. There are two underground storage sites with a working capacity of 1.3 bcm, or 28 days of consumption. The target is 35 days.

- Regulated access to storage is offered by the transmission system operator, Enagas. In LNG contracts involving third-party access, a ten-day storage service is included in the tariff. In transport TPA contracts, a five-day storage service is included in the tariff.

- New players are entering the gas market.

GAS DEMAND

- **Share of gas in TPES:** 12% (2000)

Spanish gas consumption reached 16.7 bcm in 2000.

Figure 75: Spanish gas consumption by sector in 2000

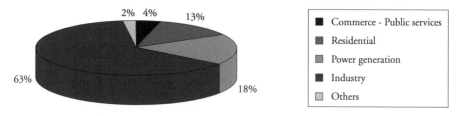

- Commerce - Public services
- Residential
- Power generation
- Industry
- Others

Temperature-sensitive sales account for a small proportion of the market, 17% in 2000. This is partly due to Spain's Mediterranean climate, but also because of the way in which the market has been developed, with an early emphasis on large industry. The share of gas in power generation is increasing.

■ **Seasonality**

The ratio between gas sales in the peak and the lowest month of the year was about 2 to 1 in 2000.

■ **Share of gas in the power-generation sector: 9% (2000)**

Table 43: Multi-fired electricity generating capacity in Spain at 31 December 2000 (GW)

Solid/Liquid	0.30
Solids/Gas	-
Liquids/Gas	3.12
Liquids/Solids/Gas	-
Total multi-fired	3.42
Total capacity*	25.49

* from combustible fuels. Public utilities only (data from autoproducers are not available).

13% of total electricity-generating capacity by fuel is multi-fired, with plants running on natural gas playing a major role. Most of the electricity produced in Spain comes from single-fired power plants run on hydropower, coal, oil and nuclear power.

■ **Interruptible contracts and fuel switching**

Interruptible contracts have been made with large consumers, who represent approximately 20% of all consumers, to increase gas supply flexibility. These customers are offered discount prices.

Much of the industrial market is served with interruptible contracts. Since gas was brought to Spain only recently, oil-tanks and firing facilities which were used before gas introduction, are in better condition than in many other countries, and this makes such contracts easier to negotiate. Furthermore, given the lack of alternative flexibility tools interruptible sales provide the required flexibility.

Combined-cycle gas turbine plants are generally supplied with firm gas.

- **Reserves:** Spain has very little gas reserves (1 bcm at the beginning of January 2001)

- **Gas supply structure:** indigenous production 1%; imports 99%, of which:
  - Algeria 60.3% (LNG 24.4%, pipe 35.9%)
  - Norway 13.4%
  - Nigeria 10.9%
  - Trinidad and Tobago 5%
  - Libya 4.6%
  - Qatar, United Arab Emirates, Belgium (spot): 1% each (approx).

- **LNG share in gas supply:** 50.6%

- **Gas supply swing**

Figure 76: Spanish monthly gas supplies

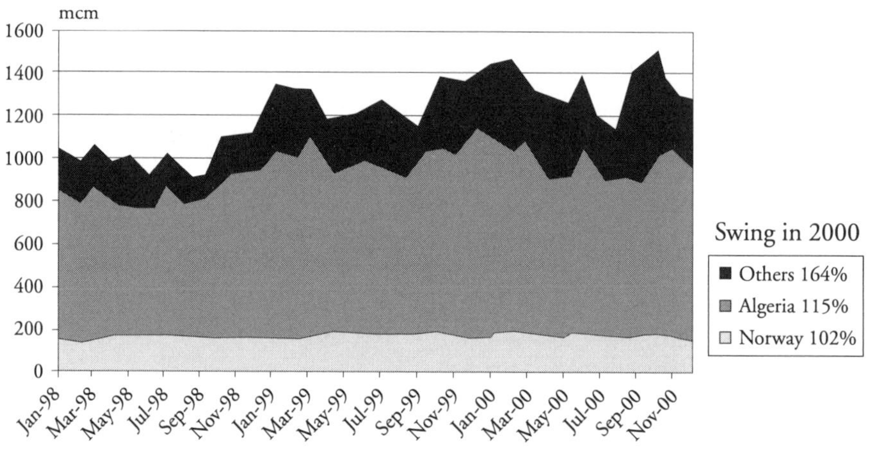

Others include imports from Belgium (spot - Interconnector), Trinidad and Tobago, Qatar, UAE, Libya and Nigeria.

Relatively little swing is provided by the suppliers. The economics of both Algerian LNG and long-distance Norwegian supplies dictate that they should deliver gas with a high load factor.

STORAGE

Storage facilities are very limited. Spain has two underground storage facilities, with a working capacity of 1.3 bcm. They represent 28 days of average consumption in 2000. Other sites are actively being sought, with studies having taken place in the Tagus and Ebro River basins, the surroundings of Madrid, Montilla (Cordova), Cantabria, Jumilla, the salt area in Alicante and in Reus (Tarragona).

Table 44: Underground gas storage in Spain

Name	Type	Operator/ Number	Working capacity (mcm)	Peak output (mcm/day)
Serrablo	Depleted gas field	Enagas	495	4.0
Gaviota	Depleted gas field	Repsol	779	5.7
Total		**2**	**1,274**	**9.7**

- **LNG import terminals**

Enagas owns the three existing LNG terminals: at Huelva, Cartagena and Barcelona. To cope with future domestic demand, three new LNG terminals have been planned, at Bilbao, Valencia and El Ferrol.

Table 45: Spanish LNG import terminals

Terminals	Storage tanks (1,000 cm of LNG)	Nominal capacity (mcm/day)	Start-up date
Barcelona (Catalogna) on the Mediterranean coast	240	29	1970
Cartagena (Murcia) on the Mediterranean coast	55 + 105[65]	10.8	1989
Huelva (Andalusia) on the Atlantic coast	160	10.8	1988

---

65 Extra capacity operational since spring 2002.

■ **Stock changes**

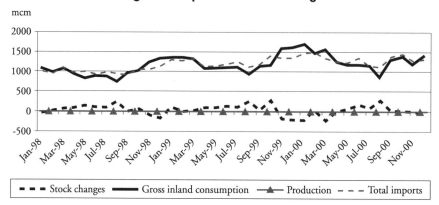

Figure 77: Spanish load balancing

Storage does not play an important role in balancing gas supply and demand. Most seasonal fluctuations in gas demand are covered by flexibility in gas imports.

REGULATORY FRAMEWORK

■ **Security of supply**

The Hydrocarbons Act of 1998 sets an indicative limit for maximum external supplies from any single country at 60%. The same limit is applied to the supplies of each individual supplier, with the exception of gas supplied to facilities with guaranteed reserves of alternative fuel.

Security of supply is enhanced through storage facilities and tanks at the LNG regasification terminals. The 1998 Hydrocarbons Act defines different operators' storage obligations:

■ Transporters who deliver gas to the system must maintain stocks equivalent to 35 days of their sales to distributors;

■ Traders must maintain stocks equivalent to 35 days of their sales;

■ Qualified consumers who buy from unauthorised traders must maintain stocks corresponding to 35 days of consumption.

A Royal Decree in 2000 created a new body, the Transmission System Operator. ENAGAS SA was named as system operator, and its capital opened up to new shareholders, with no individual company allowed to control more than 35% of its shares.

ENAGAS' role as system operator is defined as:

- controlling the short- and medium-term operation of the system, with the aim of guaranteeing continuity and security of supply;

- applying daily balancing rules for system users;

- proposing the development of transportation and storage capacity to the Minister of Economy;

- proposing emergency plans on an annual basis, taking into consideration possible disruption scenarios.

■ Access to transportation, storage and regasification terminals

Spain has opted for regulated third-party access based on published maximum tariffs. Twenty-eight companies have a gas-trading license in Spain. Not all of them have really started to operate. During 2001, the liberalised part of the market already accounted for nearly 40% of total gas sales.

SWEDEN

- Sweden has a small gas market of less than 1 bcm in 2000.
- Gas is imported from Denmark. Gas demand seasonality is covered by flexibility in imports.
- Sweden has no gas storage facility.

GAS DEMAND

- **Share of gas in TPES:** 1.4% (2000)

Natural gas currently accounts for less than 2% of total Swedish energy consumption, at 0.86 bcm. In the 26 municipalities that have access to natural gas, it accounts for 20% of energy consumption, a level equivalent to that in the rest of Europe.

Industrial plants, where gas serves both as raw material and as fuel for heating, make up 46% of the market. Power generation accounts for 30%; households for 12%. The number of final customers is approximatley 55,000, most of them are single-family houses and apartment blocks.

Figure 78: Swedish gas consumption by sector in 2000

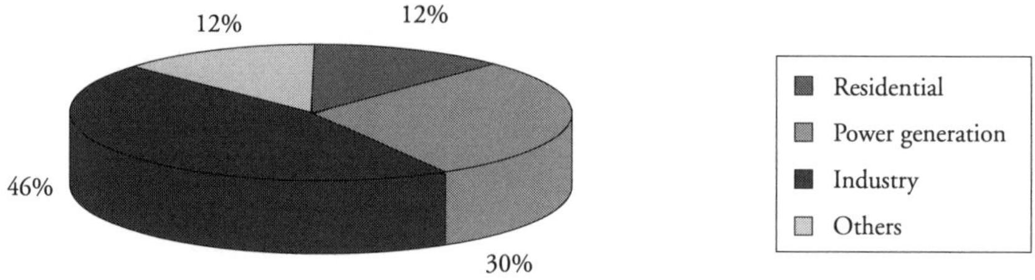

- **Seasonality**

The ratio between gas sales in the peak and the lowest month of the year was 3.5 to 1 in 2000.

- **Share of gas in the power generation sector:** < 1% (2000)

GAS SUPPLY

- **Gas supply structure:** indigenous production zero; imports 100%

All Sweden's gas comes from the Tyra field in the Danish sector of the North Sea. After transiting Denmark, a pipeline under Öresund brings the gas ashore in Sweden, at Klagshamn outside Malmö. A 300-km trunk main extends from Trelleborg in the south to Gothenburg.

- **Supply swing**

Danish imports offered a swing of 161% in 2000.

Figure 79: Swedish monthly gas imports

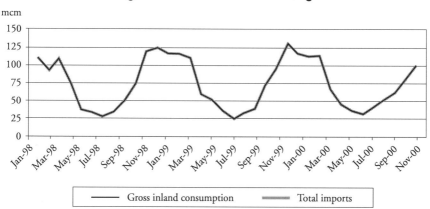

STORAGE

There is no gas storage in Sweden due to unfavourable geological conditions. However, a lined-rock cavern demonstration storage site is under construction.

- **Stock changes**

The graph below shows that seasonal variations in gas demand are covered by flexibility in import contracts.

Figure 80: Swedish load balancing

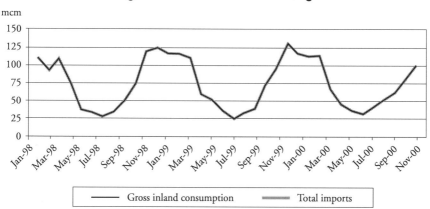

——— Gross inland consumption   ▓▓▓ Total imports

REGULATORY FRAMEWORK

- **Access to the grid**

Access to the grid is regulated.

SWITZERLAND

■ Switzerland has a small gas market which consumed 3 bcm in 2000. The temperature-sensitive market accounts for 60% of gas sales. Seasonal variations in gas demand are large.

■ Switzerland does not produce gas and has no gas storage facility.

■ The high seasonality of demand is covered by flexibility in import contracts and leased storage capacity in France.

■ Interruptible contracts play an important role.

GAS DEMAND

■ **Share of gas in TPES:** 9% (2000)

Swiss demand for gas has increased at a rate of more than 12% per year since 1973, but the share of natural gas in total energy supply remains lower than the average of IEA European members.

Gas consumption reached 3 bcm in 2000. Almost 60% of natural gas is consumed by the residential and commercial sectors. Demand seasonality is therefore high. Industrial consumption accounts for 31% of total sales.

Figure 81: Swiss Gas Consumption by Sector in 2000

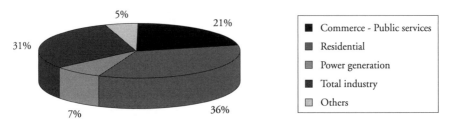

- ■ Commerce - Public services
- ■ Residential
- ■ Power generation
- ■ Total industry
- □ Others

■ **Seasonality**

The ratio between gas sales in the peak and the lowest month of the year was 4 to 1 in 2000.

- **Share of gas in the power generation sector:** 2% (2000)

Switzerland has no multi-fired power plants. Most of the electricity is generated from hydro-power plants.

- **Interruptibles and fuel switching**

Forty-five percent of total gas consumption is under an interruptible basis (1999). Thirty-two percent of gas demand for industrial customers and 100% for the power and heat generation sector are on an interruptible basis. Deliveries to interruptible customers can be interrupted when temperatures drop below minus 10°C.

Switzerland stores heating oil as a backup for interruptible gas supply contracts. Light heating-oil stocks have been built up to cover approximately four-and-a-half months of natural gas consumption.

GAS SUPPLY

- **Gas supply structure:** indigenous production zero; imports 100%, of which:
  - France 11.5%
  - Germany 56.5%
  - Netherlands 19.1%
  - Russia 11.2%
  - Italy 1.7%

Source: Swissgas.

- **Supply swing**

Figure 82: Swiss monthly gas imports

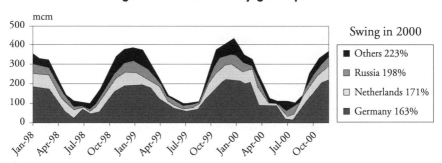

Others include imports from France and Italy.

As can be seen from the graph, suppliers offer a great deal of flexibility.

STORAGE

Switzerland has no gas storage on its territory. Gaznat has an agreement with Gaz de France that allows the Swiss company to withdraw a limited amount of gas from the French storage facility of Etrez, northwest of Geneva.

■ **Stock changes**

Figure 83: Swiss load balancing

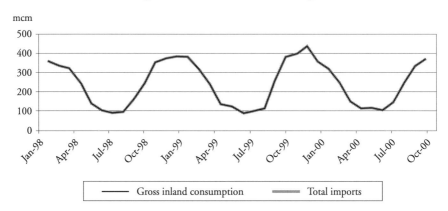

Seasonal demand is covered by flexibility in contracts and storage capacity rented in France. Interruptible contracts are concluded with large customers.

REGULATORY FRAMEWORK

■ **Security of supply**

Increasing interconnections with the rest of Europe have improved Switzerland's security of supply. Since the expansion of the Transitgas system, Switzerland has become an important transit country for gas from Northern Europe to Italy.

■ **Access to the transmission network**

As a non-EU member, Switzerland has no legal obligation to adopt the EU Gas Directive. However, because of its comprehensive integration into the European natural-gas grid system and the liberalisation processes under way in other sectors, the Swiss gas market is going to be opened up. New arrangements will take into account both European trends and Swiss requirements. TPA to the high-pressure grid has been provided by legislation since many years.

TURKEY

■ Gas consumption reached 14.8 bcm in 2000. Turkey has very rapid growth in gas demand.

■ Natural gas is mainly used by the power and residential sectors (83%).

■ Demand seasonality is relatively high and is entirely met by import flexibility.

■ Turkey does not yet have any underground gas storage facility on its territory, but it does have an LNG receiving terminal.

GAS DEMAND

■ **Share of gas in TPES:** 16% (2000)

■ **Gas demand structure**

Turkish gas consumption reached 14.8 bcm in 2000. With 9.3 bcm in 2000, power generation accounted for 64% of total gas demand, followed by the residential sector with 3.2 bcm, and industry with 2.1 bcm. In most consuming sectors, gas has replaced oil and coal. The government has encouraged the use of natural gas to replace lignite in the residential sector to reduce urban pollution.

Gas demand is expected to triple between 2000 and 2010.

Figure 84: Turkish gas consumption by sector in 2000

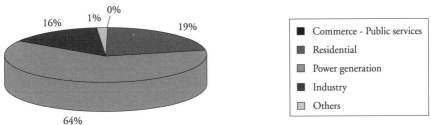

- **Seasonality**

The ratio between gas sales in the peak and low month of the year was 1.5 to 1 in 2000. Seasonality of gas demand by sector is indicated in the graph below:

Figure 85: Turkish monthly gas consumption by sector

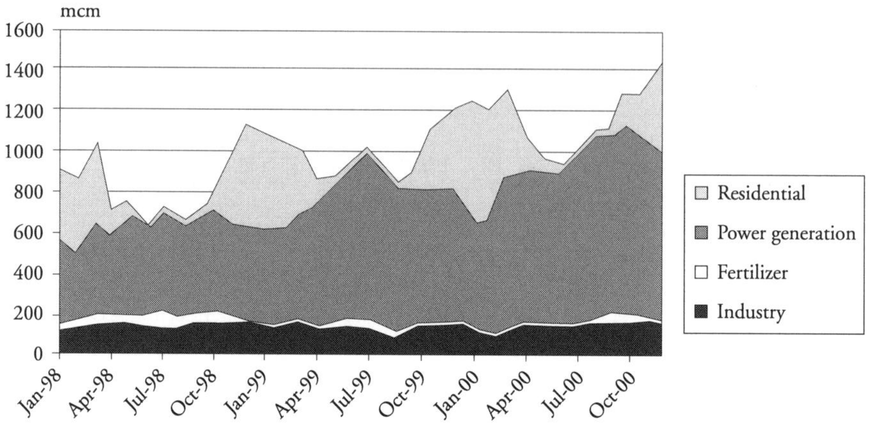

Source: Ministry of Energy and Natural Resources of Turkey.

The seasonality of gas demand by residential consumers is very large. Sales in this sector are concentrated in the winter months (73% of annual sales between December and March).

- **Share of gas in the power generation sector: 36% (2000)**

Table 46: Multi-fired electricity generating capacity in Turkey at 31 December 2000 (GW)

Solid/Liquid	0.41
Solids/Gas	0
Liquids/Gas	2.14
Liquids/Solids/Gas	0
Total multi-fired	2.55
Total capacity*	16.06

* from combustible fuels.

Only 16% of total electricity generating capacity by fuel is multi-fired, with plants running on natural gas playing a major role. Most electricity produced in Turkey comes from single-fired power plants run on hydropower and coal. This situation, however, is rapidly changing with an expected tripling of gas consumption in electricity plants by 2010.

■ **Interruptibles and fuel switching**

BOTAS has used interruptible customers as a tool to reduce demand at times of peak gas use. The increase in natural gas consumption in cities during winter leads to an interruption of gas supply to interruptible customers for peak shaving purposes. Interruption in gas supply is effected with an eight hours prior notice. The notice period can be shorter in special circumstances. The interruptions are usually made between December and March and can last for several days or weeks. The price for interruptible customers is on average 12% less than the price for non-interruptible customers.

GAS SUPPLY

- ■ **Reserves:** 9 bcm   **R/P:** 14 years
- ■ **Gas supply stucture:** indigenous production 4%; imports 96%, of which:
   - Russia 65%
   - Algeria 26%
   - Nigeria 5%

Supplies from Algeria and Nigeria are imported in the form of LNG. Turkey has begun to top up its long-term contracts with spot deliveries.

In order to secure gas supplies, BOTAS signed eight natural gas purchase agreements with six different countries for 67.8 bcm per year of natural gas or LNG imports. These break down into 4 bcm per year of LNG from Algeria, 1.2 bcm of LNG from Nigeria, 14 bcm of natural gas from Russia via the West, 16 bcm of natural gas from Russia through the Black Sea, 10 bcm of natural gas from Iran, 16 bcm of natural gas from Turkmenistan and 6.6 bcm of natural gas from Azerbaijan.

- ■ **LNG share in supply:** 31%

■ Supply swing

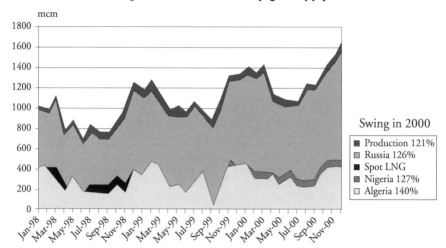

Figure 86: Turkish monthly gas supply

Source: Ministry of Energy and Natural Resources of Turkey.

The graph indicates the increasing amount of gas imported from Russia, Algeria and Nigeria. Little flexibility is offered by suppliers.

STORAGE

Turkey has not yet built any underground gas storage. However, studies are proceeding on the possibility of storing gas in underground storage facilities.

Within the framework of this project, the Salt Lake Natural Gas Underground Storage Project will use the salt domes in Salt Lake. Engineering and consultancy studies are under way for the project. A call for construction tenders will be launched in late 2002. Environmental impact assessment studies of the project are also being carried out.

The Tarsus Natural Gas Underground Storage Project will use the sodium carbonate beds of Sisecam Soda San. A.S., in Mersin, as underground storage facilities. Pre-feasibility studies are being carried out.

Studies are also underway on using TPAO's Northern Marmara and Degirmenköy gas fields as underground gas storage facilities following their depletion. Total storage capacity of these two facilities will be 1.6 bcm, and they are to become operational in 2005.

■ LNG import terminals

Table 47: Turkish LNG import terminals

Terminals	Storage tanks (1,000 cm of LNG)	Nominal capacity (mcm/day)	Start-up date
Marmara Erelisi	255	13	1994

■ Stock changes

Figure 87: Turkish load balancing

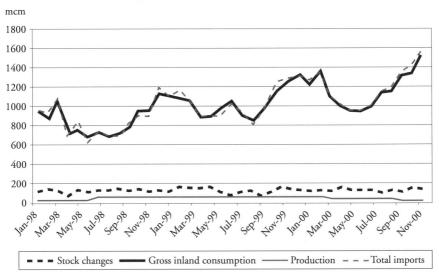

Demand seasonality, 120% in 2000, can be met with flexibility in imports.

REGULATORY FRAMEWORK

■ Security of supply

In its monopoly role, BOTAS has been responsible for securing the uninterrupted delivery of natural gas to its customers. To maintain both strategic and technical security, BOTAS has diversified its gas supply sources by country, its importing pipeline routes and the way gas is imported into the country (by pipeline and LNG). BOTAS is now building storage facilities.

BOTAS applies international standards for gas-system construction and operation. Under the new Natural Gas Market Law of May 2001, transmission and storage companies have an obligation to prove to the Natural Gas Market Regulatory Agency that their services are economic and safe.

The new law obliges gas importers and wholesalers to cover 10% of their imported gas with gas storage. The companies have five years to comply.

■ **Access to the transmission grid**

Since the adoption of the Natural Gas Market Law on 2 May 2001, the regulation of Turkey's gas industry has been changing. The new law foresees the controlled opening of the gas industry to competition, with the aim, among others, of harmonising Turkish legislation with EU law. The Act allows a twelve-month transition period. The Council of Ministers can extend the transition by six months, but only once.

Restructuring of the gas market is also intended.

- Natural gas supply, transmission and distribution are to be unbundled. BOTAS is to be split into two units, one responsible for trading, which is to be privatized later, the other for transmission, which is to be kept as a State Economic Enterprise. The two distributors owned by BOTAS, in Bursa and Eskisehir, are to be corporatised and then privatised.

- BOTAS will continue to own and operate the national transmission network, as well as LNG and storage facilities. It will offer services under a system of non-discriminatory, regulated and published prices and access conditions. These prices and access conditions are to be regulated by a new regulatory agency.

Network tariffs are based mainly on distance and volume. Storage tariffs are freely negotiated between storage companies and users.

UNITED KINGDOM

- The British gas market has grown strongly in the past decade. Gas consumption now amounts to more than 100 bcm.

- The temperature-sensitive sectors (residential/commercial) represent 42% of gas sales. But the ratio between gas sales in the peak and the lowest month of the year is only about two to one. This is due to the high share of gas in the power sector, mainly used for base load.

- Interruptible contracts play an important role in balancing supply and demand.

- In the past, the very high production swing offered by North Sea producers limited storage needs. This is changing, with increasingly high load factors from North Sea producers.

- Since October 1998, the Interconnector has played a role in rebalancing UK gas supply and demand.

- UK has nine storage sites (of which five are LNG peak shavers). Peak sendout is 138 mcm/day and the working capacity is 3.6 bcm, or 4% of annual gas consumption.

- Access to storage is deregulated, but subject to some conditions. In July 2001, US Dynegy bought BG Storage.

- New independent storage projects are now operational and there are also "virtual storage" services, as well as a secondary market for storage.

GAS DEMAND

- **Share of gas in TPES: 38% (2000)**

Gas demand has grown rapidly over the last ten years to 102 bcm in 2000. The power sector accounted for 29% of gas demand in 2000, an increase of 10 percentage points over 1996. This share is expected to continue to rise with the removal of the gas moratorium which prevented the building of new gas-fired power plants. The temperature-sensitive sectors represent 42% of gas sales.

Figure 88: UK gas consumption by sector in 2000

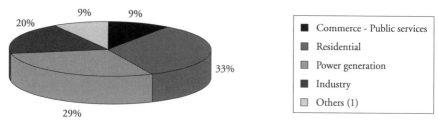

9%   9%
20%
33%
29%

■ Commerce - Public services
■ Residential
■ Power generation
■ Industry
☐ Others (1)

(1) The energy sector accounts for 70% of the "others" category.

UK gas consumption hit a new one-day record on 3 January 2002, when demand reached 435 mcm, beating the previous record of 427 mcm.

■ **Seasonality**

The ratio between gas sales in the peak and the lowest month of the year was 2 to 1 in 2000.

Figure 89: UK quarterly gas demand by sector

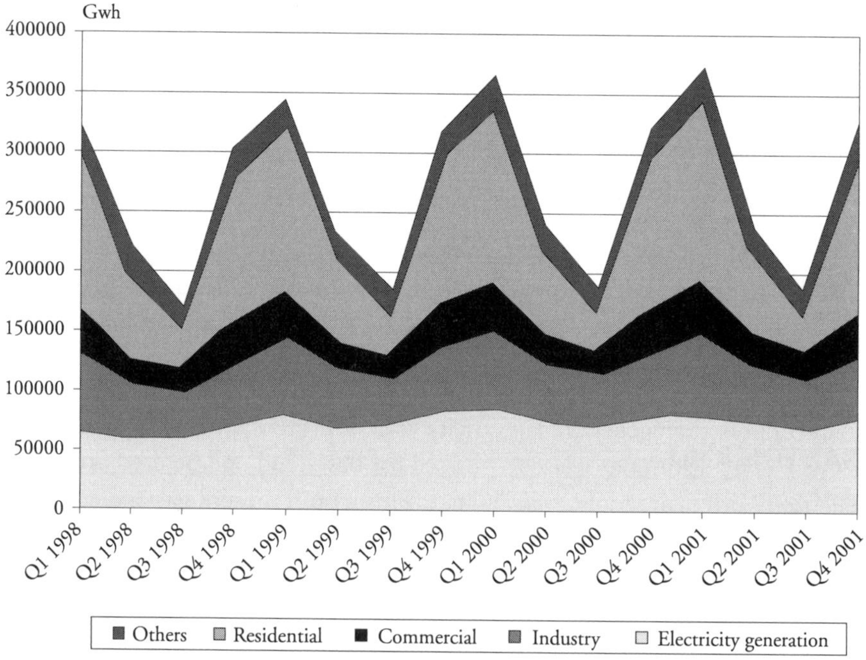

Gwh

400000
350000
300000
250000
200000
150000
100000
50000
0

Q1 1998 Q2 1998 Q3 1998 Q4 1998 Q1 1999 Q2 1999 Q3 1999 Q4 1999 Q1 2000 Q2 2000 Q3 2000 Q4 2000 Q1 2001 Q2 2001 Q3 2001 Q4 2001

■ Others   ☐ Residential   ■ Commercial   ■ Industry   ☐ Electricity generation

Source: DTI.

DTI gives only quarterly breakdowns of gas consumption by sector. The graph nevertheless shows the seasonal variations in demand in the residential sector.

In this sector, sales during the first quarter of 2001 were 4.5 times higher that sales in the third quarter.

■   Share of gas in the power generation sector: 39% (2000)

Table 48: Multi-fired electricity generating capacity in the United Kingdom at 31 December 2000 (GW)

Solid/Liquid	7.49
Solids/Gas	0
Liquids/Gas	0.76
Liquids/Solids/Gas	0
Total multi-fired	8.25
Total capacity*	61.96

* from combustible fuels.

Only 13% of total electricity-generating capacity by fuel is multi-fired, with plants running on natural gas having a marginal role; most of electricity produced in UK comes from nuclear plants and single-fuel fired coal and gas plants.

Combined-cycle gas turbine generation has grown very quickly in the United Kingdom. The share of gas in the power-generation sector was negligible 10 years ago. It is now 39%.

Most of CCGT generation tends to run as base load. This is because gas has been relatively cheap and the arbitrage opportunities between gas and power have gone only one way - favouring the burning of gas to generate electricity. Now, with the cost of gas rising and that of electricity decreasing, the arbitrage opportunities are no longer necessarily one way. When electricity prices are low, a generator with committed gas may do better by selling the gas on the spot market rather than generating electricity with it. If this happens frequently, some CCGT generation may cease to be used for base-load and be used only when electricity prices are high, at peak load. This would tie spot gas and electricity prices more closely together. It would also have implications for flows of gas across the Interconnector, which could switch to being a swing provider in response to variation in UK spot-gas prices.

In 2001, gas use for electricity generation was 2% lower than in 2000, although two new gas-fired power stations started to generate during 2001 and 4 others made their first full year contributions. High gas prices led to some stations generating for fewer hours than they would have liked.

■ Interruptible contracts and fuel switching

Sales of gas on an interruptible basis accounted for about 24% of total UK gas sales in 1998 and 1999, rising to around 26% in 2000, but fell to 23% in 2001 largely because of the high gas prices. In 2000, interruptible sales to the electricity generation sector are estimated to have accounted for 69% of total sales to that sector (61% in 2001).

The prevalence of interruptible sales to the power sector seems at first glance inconsistent with the reduced capacity of gas-fired power plants able to switch to other fuels. However, power producers are able to cut off their electricity generation from gas thanks to the existing over-capacity in the overall electricity generation system.

GAS SUPPLY

- **Reserves:** 760 bcm   **R/P:** 7 years

- **Gas supply structure:** indigenous production 98%; imports 2%, of which:

  - Norway 1%

  - Interconnector 1%

UK is a net exporter of gas through the interconnectors to Continental Europe and Ireland. In 2000, UK exported 13.38 bcm.

■ Gas production

Production was 115.3 bcm in 2000. Market liberalisation has given producers a strong incentive to explore for, appraise and develop new fields on a "just-in-time" basis and to accelerate production from fields as they are brought on stream.

As in the Netherlands, a key feature of the UK gas production was the high swing offered by offshore producers, which averaged 160% up to 1995. However, as the cost of developing incremental sources of production has increased, the swing has declined. It was 124% only in 2000.

Although the production swing in general has decreased, some fields still do offer high swing. This is the case of the Morecambe field, owned by Centrica. The field offers a great deal of flexibility (swing production of 167% in 2001), and it helps Centrica to meet seasonal variations in its customers' gas demand.

The further development potential of the UK Continental Shelf lies mainly in small fields located near existing infrastructure in the southern and central North Sea. This means less flexibility.

Trends over the past five years indicate a levelling off of total UK gas reserves and increased production. It is estimated that UK indigenous supplies will tighten by 2005 and that the UK will have to import gas regularly by 2005. It will need seasonal top-up deliveries even earlier.

Figure 90: UK monthly gas production

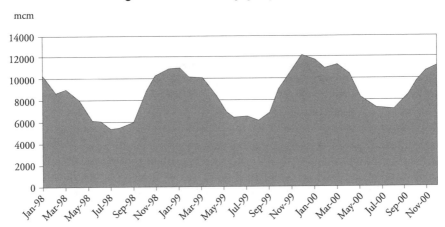

- **Imports flexibility**

Britain's gas market ceased to be isolated from mainland Europe in October 1998, when the UK-Belgium Interconnector entered service. The Interconnector[66], a 230-km pipeline linking Bacton to Zeebrugge, can import gas to the UK as well as export it, although importing capacity is currently only 40% of export capacity, which is 20 bcm/year. Since it started operation in October 1998, the Interconnector has had a major impact on both imports from and exports into the UK. Approximately 10 bcm per year have been sold to Continental Europe under long-term (10-15 years) contracts. Depending on price differentials, extra volumes may be sold to or bought from the Continent. If gas is sold to the UK the volumes would first net with the volumes under the long-term contracts and eventually lead to a net flow into UK. This happened several times in 2000 and 2001, as prices were higher on the NBP

---

66   Source : UK Trade in Natural Gas, Energy Trends, DTI, Fevruary 2001.

than in Zeebrugge. The Interconnector thus plays an important role for balancing UK supply and demand. Since the facility opened, except for 1999, the UK has been a net importer of seasonal gas every winter. The amount imported is determined by a combination of temperatures, production and storage availability in the UK and the cost of European gas.

Figure 91: UK monthly gas imports

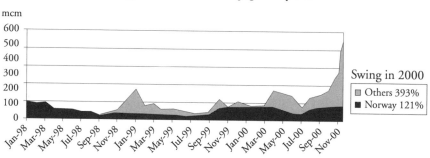

In 2000, large volumes of gas were sold for export to Europe to take advantage of higher European gas prices. Higher oil prices prompted Continental buyers to maximise their sourcing of cheaper UK supply. Gas exports from Britain through the Interconnector were often close to their 1.93 bcfd forward-flow capacity.

In 2001, gas imports were 17% higher than in 2000 and gas exports 5.5% lower. This was primarily due to a high level of imports from Belgium during the early months of 2001 and also high levels of imports in the fourth quarter of 2001, as suppliers took advantage of cheaper gas from the Continent rather than indigenous UK production. First production from the Norwegian Vesterled pipeline system in October 2001 also raised the amount of gas imported.

STORAGE

In the past, the very high production swing offered by North Sea producers limited UK storage needs, and these were adequately covered by five LNG peak-shaving units. The situation has changed, with higher load factors from North Sea producers. Now the UK has nine storage sites (including the five LNG units). Peak sendout is 138 million cm/day and the working capacity is 3.6 bcm (4% of annual consumption). Two independent storage, at Hatfield Moors and Holehouse Farm, are now operational and other sites are planned at Humbly Grove, Warmington and Aldbrough.

Table 49: Underground gas storage in the United Kingdom

Name	Type	Operator/ Number	Working capacity (mcm)	Peak output (mcm/day)
Hornsea	Salt cavity	Dynegy*	325	18.2
5 LNG Peak Shaving	Peak shaving	Transco	374	75
Rough	Semi depleted gas field	Dynegy*	2,800	42
Hatfield Moors	Salt cavity	Scottish Power	116	1.7
Holehouse Farm	Salt cavity	Aquila	12	1.5
Total		**5**	**3,627**	**138.4**

* Until 2001, most storage sites were owned by BG Storage, a division of BG Group. In July 2001, US Dynegy bought BG Storage for £421 million ($590 million). The sale included the Rough offshore facility in the southern North Sea, the Easington gas processing terminal and nine salt cavities in Hornsea, east Yorkshire. The deal was approved by regulatory offices in November 2001.

■ **LNG import terminal**

The LNG receiving terminal (Canvey Island) is nowadays closed. There are plans to build new regasification terminals at Milford Haven and on the Isle of Grain.

■ **Stock changes**

Figure 92: UK load balancing

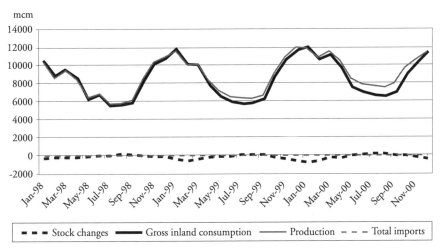

As can be seen from the graph, storage plays a relatively minor role in balancing supply and demand. Most of fluctuations in gas demand are covered by production swing.

REGULATORY FRAMEWORK

- **Security of supply**

Key obligations on network operators, shippers and suppliers are set out in their operating license conditions[67].

- *Suppliers* are required to ensure that they can fulfil their peak demand from residential consumers for gas on the coldest day expected in a period of 20 years and aggregate daily demand over the coldest winter or year expected in a period of 50 years.

- *Shippers* are required to book sufficient capacity in the gas pipeline system to meet the peak aggregate demand of its customers for gas on the coldest day expected in a period of 20 years.

- *Transmission system operators'* primary obligation is to install sufficient delivery capacity to meet all non-interruptible capacity demand on the coldest day expected in a period of 20 years. They have to report annually to the regulator Ofgem on their performance in terms of availability, security and quality of supply and have statutory duties to develop and maintain the system. They also have an obligation to provide incentives to suppliers to ensure that the supply security standards for residential customers are met. These obligations are the responsibility of Transco as the national transmission system operator.

In addition, Ofgem can set certain standards of performance for gas transportation, supply and metering. There are penalties for the breach of licence conditions and of the standards set by the regulator.

The Energy Act 1976 gives statutory powers to the government to deal with emergencies in which supplies of oil, electricity and gas are disrupted. In the case of gas, Transco would act as the national emergency co-ordinator. If a supply disruption were unavoidable, supply to households and other priority gas consumers, such as hospitals would be given priority, but these rules have

---

67 DTI Standard License Conditions, September 2001 : http ://www.dti.gov.uk/energy/gas-electricity.htm

not been invoked since liberalisation and third-party competition were introduced.

■ Access to transmission network and storage

Access is regulated on the basis of published tariffs. The network is 100% open. Transco is the owner of the majority of Britain's gas transportation system. It receives gas from several coastal reception terminals around Great Britain, and transports it to more than 20 million industrial, commercial and domestic customers. Its network is made up of around 275,000 km of pipeline, consisting of high-pressure national and regional transmission systems, and lower-pressure distribution systems. The interconnector to Belgium links Transco's gas transportation system to Continental Europe's high-pressure gas grid. A second interconnector supplies gas to Ireland and Northern Ireland.

Transportation is provided according to the terms of a standard contract applicable equally to all prospective users, and known as the Network Code. The transporter is obliged to meet all reasonable requests for service as part of its common-carriage obligations. A refusal to do so is appealable to the regulator.

Dynegy provides storage on a non-discriminatory basis to any shipper. Storage capacity is auctioned subject to reserve prices.

Transco owns, operates and develops all LNG peak-shaving facilities in Great Britain.

UNITED STATES

- The United States have the biggest gas market in the world. The residential and commercial sectors represent 36% of gas consumption.

- The seasonality of gas demand is large in the residential sector. Sales during December 2000 were 7.4 times higher than in August.

- The US is 84% self-sufficient in gas.

- Gas storage is highly developed. Total working capacity amounts to 110 bcm, or 17% of total gas consumption. Storage is used to meet seasonal and peak fluctuations in gas demand.

- LNG import terminals and LNG peak shaving-units help smooth peak daily gas demand.

- Roughly 10 to 15% of all natural gas deliveries to US consumers by interstate pipelines in 1997 were on an interruptible basis, down substantially from roughly half of all deliveries in the late 1980s.

- The US has a complex and extensive pipeline infrastructure. Utilisation of pipelines in parts of the west have recently been well above 95% on a continuing basis.

GAS DEMAND

- **Share of gas in TPES:** 24% (2000)

The United States have the biggest gas market in the world, with 638.6 bcm consumed in 2000.

Figure 93: US gas consumption by sector 2000

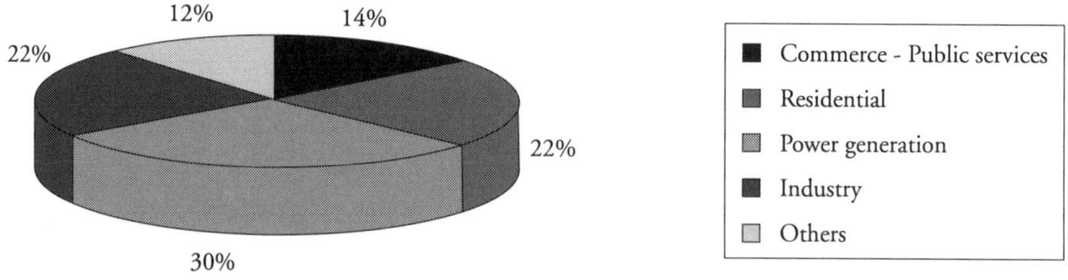

12%   14%

22%

22%

30%

- Commerce - Public services
- Residential
- Power generation
- Industry
- Others

The power generation sector (including CHP) was the largest user of natural gas in 2000. Natural gas is the largest energy source in the residential sector and the fastest-growing energy source for electricity generation.

■ **Seasonality**

The ratio between gas sales in the peak and the lowest month of the year was approximately 1.8 to 1 in 2000. The seasonality of gas demand is important in the residential sector. Deliveries in December 2000, the peak month of the year, were 7.4 times higher than in August, the lowest month.

Figure 94: US gas demand seasonality

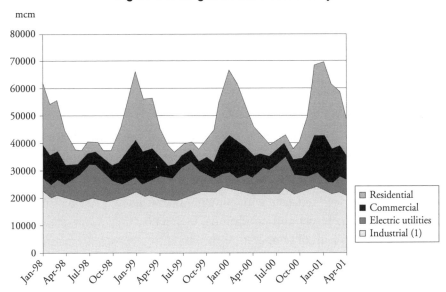

(1) Includes CHP.
Source: Energy Information Administration.

■ **Share of gas in the power generation sector:** 16% (2000)

The natural-gas share of US electricity generation, including cogeneration, rose from 13.2% in 1996 to about 16% in 2000 (EIA). When cogeneration is excluded, the share was 13% in 2000.

Natural gas for electricity generation is mainly used during peak demand periods in summer times, and it is the preferred energy source for new generating capacity. About 90% of planned additions to electricity-generation capacity over the next few years are designed to use natural gas as a primary fuel source.

Table 50: Multi-fired electricity-generating capacity in the United States at 31 December 2000 (GW)

Solid/Liquid	6.29
Solids/Gas	38.65
Liquids/Gas	56.56
Liquids/Solids/Gas	164.99
Total multi-fired	266.49
Total capacity*	629.97

* from combustible fuels.

42% of total electricity-generating capacity by fuel is multi-fired, with plants running on natural gas having a large role. Dual gas and heavy fuel oil capacity in the US stands at about 60 GW.

■ **Interruptibles and fuel switching**

Roughly 10 to 15% of all natural gas delivered to US consumers by interstate pipelines in 1997 were on an interruptible basis, down substantially from roughly half in the late 1980s.

Interruptible service contracts with pipeline operators or local distribution comapnies vary in terms and conditions. Generally, they allow for service interruptions as a result of temperature-threshold triggers or system operating conditions (for example, when line pressure is threatened by very high draws on the system). In addition, some contracts provide firm service only for a limited duration, usually a month, or on a seasonal basis, with suspensions of service permitted during the winter. Suspension of service is not considered an interruption so long as the terms of the contract are fully met.

In 1998 (the latest year with available data), total sales on an interruptible basis amounted to 140 bcm (25% of total gas sales). Of total sales to the industrial sector, 38% were on interruptible basis (93 bcm), as were 36% of sales to electric utilities (33 bcm) and 15% of sales to commercial consumers (13 bcm).

GAS SUPPLY

■ **Reserves:** 5,024 bcm   **R/P:** 9 years

- **Gas supply structure:** indigenous production 84%; imports 16%, of which:
  - Canada 15%
  - Mexico, Trinidad, Oman, Qatar, UAE, Algeria, Nigeria 1%

- **LNG share in gas supply:** 1%

- **Gas production**

US production amounted to 540 bcm in 2000. Production remains roughly constant year round, swing production was 107% in 2000. Production is concentrated in the southern and central US States.

Figure 95: US monthly gas production

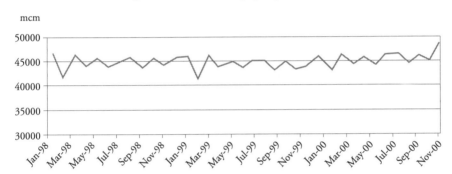

- **Gas imports**

US is largely self-sufficient in gas, with 84% coming from domestic production, although imports in the form of LNG picked up in 2000 and 2001 due to high

Figure 96: US monthly gas imports

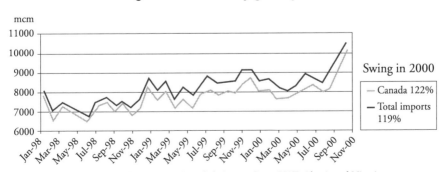

Total imports include imports from Mexico, Trinidad, Oman, Qatar, UAE, Algeria and Nigeria.

gas prices. In 2000, total gas imports amounted to 108 bcm, of which 94% came from Canada. In line with an increase of total imports, the graph below shows the increasing load supply.

■ Transportation

The US has a complex and extensive pipeline infrastructure for transporting gas from production areas to ultimate consumers. More than 165 companies operate about 278,000 miles of interstate and intrastate transmission lines, hundreds of compressor stations and numerous storage facilities, allowing gas delivery throughout the lower 48 States. In addition, more than 1,300 local distribution companies provide local delivery services through another 700,000 + miles of pipeline infrastructure. In 2000, these lines transported an estimated 22.8 tcf (645 bcm) of natural gas from supply sources to end-use markets.

Since 1990, the gas pipeline network has grown substantially, with a 27% increase in interregional capacity. The network has also become more interconnected, its routing more complex, and business operations more efficient. Except during periods of extreme weather conditions or disruptions caused by isolated pipeline outages, there has been no sustained disruption of the network since the mid-1970s.

Nonetheless, increasing growth in natural gas demand over the past several years has led to an increase in the utilisation of pipelines and has resulted in some pressure for expansion in several areas of the country. Pipeline utilisation in parts of the west, notably pipelines delivering gas to the California market, has recently been well above 95% on a continued basis. Further increases in demand could cause capacity bottlenecks to develop.

Over the past two years, 63 natural gas-pipeline construction projects (35 in 1999 and 28 in 2000) have been completed and placed in service in the United States, adding more than 12.3 bcfd of new pipeline capacity, an increase of 15% over 1998.

A major growth area in pipeline expansion during the past several years has been the export/import market for natural gas. Much pipeline construction has been to expand import capacity for Canadian gas into the U.S Midwest and Northeast. Natural gas export capacity to Mexico has more than doubled since 1996 to 2.1 bcfd.

There are 88 pipeline projects announced for the next several years totalling an additional 20.8 bcfd of capacity.

STORAGE

Gas storage is highly developed in the United States. Total working capacity amounted to 110 bcm in 2000 (or 17% of total gas consumption). On 1 November 2000, working gas-storage volumes - at 2,699 bcf or 76.4 bcm - were the lowest for the start of a heating season since 1976 and 6 bcm below the five-year average. The situation reversed itself completely in 2001 and working gas in storage is now well above the five-year average.

Table 51: Underground gas storage in the United States (2000)*

Region	Type	Number	Working capacity (mcm)	Peak output (mcm/day)
East	Aquifer	33	9,939	211.2
	Depleted gas/oil field	243	47,855	903.0
	Salt cavern	4	113	8.4
Total East		**280**	**57,907**	**1,122.6**
West	Aquifer	6	1,104	33.3
	Depleted gas/oil field	31	16,707	244.1
	Salt cavern	0	0	0
Total West		**37**	**17,811**	**277.4**
Producing regions	Aquifer	1	28	0.3
	Depleted gas/oil field	74	30,837	486.1
	Salt cavern	23	3,823	314.8
Total producing regions		**98**	**34,688**	**801.2**
Total US	**Aquifer**	**40**	**11,072**	**244.8**
	Depleted gas/oil field	**348**	**95,399**	**1,633.1**
	Salt cavern	**27**	**3,936**	**323.3**
Total		**415**	**110,406**	**2,201.2**

* Does not include LNG peak-shaving facilities (99 sites, 2.6 bcm of capacity and 329.4 mcm/day of deliverability).
Source: Energy Information Administration.

■ **LNG import terminals**

LNG imports rose to 6.6 bcm, or 1% of the total gas supply, in 2000. The two existing LNG receiving terminals, at Lake Charles and Everest, are increasing their throughput. The other two at Cove Point and Elba Island, which have been mothballed for many years are being brought back into operation, and even planned to be enlarged.

Table 52: US LNG import terminals

Terminals	Storage tanks (1,000 cm of LNG)	Sendout capacity (mcm/day)	Start-up date
Everett	99.109	12.7	1971
Lake Charles	178.396	19.8	1982
Cove Point (mothballed)	141.584	28.3	1978
Elba Island (mothballed)	116.099	12.5	1978
Total	**441**	**78.3**	

Source: Energy Information Administration.

■ **Stock changes**

Figure 97: US load balancing

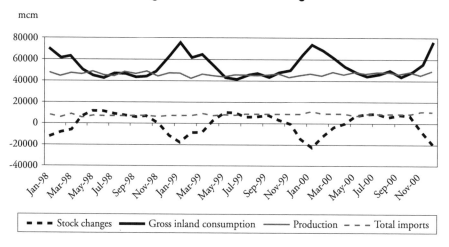

This graph demonstrates the key role of underground storage in balancing gas supply and demand, with storage facilities being filled in summer months and drawn down in the winter.

- **Security of supply**

According to the new National Energy Policy published in May 2001 moderating the recurrence and severity of "boom and bust" cycles while meeting increasing demand at reasonable prices is one of the major challenges facing the US natural gas industry today. At the core of the policy are proposals to ensure adequate domestic energy supply and infrastructure.

The National Energy Policy includes a set of recommendations to enhance oil and gas exploration and production and investments in gas pipelines by:

- promoting enhanced oil and gas recovery from existing wells through new technology;

- improving oil and gas exploration technology through continued partnership with public and private entities;

- reviewing land status and impediments to federal oil and gas leasing;

- expediting the ongoing study of impediments to federal oil and gas exploration and development;

- reviewing public lands withdrawals and lease stipulations;

- considering economic incentives for oil and gas development such as royalty reductions for enhanced oil and gas recovery; for reduction of the risk associated with production in frontier areas or deep gas formations, and for development of small fields that would otherwise be uneconomic;

- reviewing the regulation of energy-related activities and the siting of energy facilities in the coastal zone and on the Outer Continental Shelf, continuing Outer Continental Shelf oil and gas leasing and approving of exploration and development plans on predictable schedules;

- considering further lease sales in the National Petroleum Reserve - Alaska, including areas not currently leased within the northeast corner of the reserve;

- authorising exploration and possible development of the 1002 Area of the Arctic National Wildlife Refuge;

- expediting construction of a pipeline to deliver natural gas to the lower 48 states;

- supporting legislation to improve the safety of natural gas pipelines;

- continuing efforts to improve pipeline safety and expediting pipeline permissions;

- considering improvements in the regulatory process governing approval of interstate natural-gas pipeline projects.

■ Access to the transmission system

The US natural gas market today is extremely open and competitive. Wellhead gas prices were deregulated between 1979 and 1989 and are now subject to market forces.

Interstate pipelines are regulated by the Federal Energy Regulatory Commission, which regulates pipeline rates, construction of new or expanded pipelines and facilities and certain environmental aspects. The Commission ensures open, non-discriminatory access to gas transport for all competing suppliers.

Commission's Order 436 issued in 1985 provided for open access to pipelines by requiring them to transport third-party gas. FERC Order 636, issued in 1992, unbundled pipeline sales and transportation functions, with transportation remaining a regulated monopoly but with sales opened to competition. Partly as a consequence, there are now some 260 unregulated independent natural gas marketers. The system of pipelines on which producers compete is extensive, with 278,000 miles of pipe.

Natural gas distribution is regulated by state public utility commissions that are responsible for regulating all aspects of gas distribution including consumer rates. Many states are opening up functions like billing and metering to competition, while continuing to regulate local grids. There are some 1,400 local gas-distribution utilities in all, varying in size from small companies with a few thousand customers to several that have over a million customers.

GLOSSARY OF TERMS AND ABBREVIATIONS

Associated gas - Natural gas found in a crude oil reservoir, either separate from or in solution with the oil.

At the beach - (UK) When gas has been brought ashore to a terminal by producers but is not yet in the national transmission system, the gas is called at the beach.

Balancing mechanism - In a natural gas pipeline network, the means of ensuring that supply does not outstrip demand, or vice versa.

Base-load - The minimum amount of electric power delivered or required over a given period of time at a steady rate.

bbl - barrel.

bcf - billion cubic feet.

bcf/d - billion cubic feet per day.

bcm - billion cubic metres.

Beach/border price - Price of gas delivered to the beach or border terminal.

Btu - British thermal unit.

Bundled services - Two or more gas services provided at a combined charge - e.g. gas transportation and storage.

Calorific Value (CV) - A measure of the amount of energy released as heat when a fuel is burned. It may be measured gross or net, where gross includes the heat produced when the water vapour is condensed into a liquid and net does not.

Churn - The ratio of traded volumes at a hub to actual physical volumes.

Combined-cycle gas turbine (CCGT) - An energy-efficient gas turbine system where the first turbine generates electricity from the gas produced during fuel combustion. The hot gases pass through a boiler and then into the atmosphere. The steam from the boiler drives the second electricity generating turbine.

Combined heat and power (CHP) - A power station system that uses both gas and the heat/steam generated to produce electricity. Also known as cogeneration.

Compressor station - Gas loses pressure as it travels over long distances. A compressor station, usually a gas turbine engine, is an installation which recompresses the gas to the required pressure.

Cost, Insurance and Freight (cif) - A cif price means that the cost of cargo, insurance and travel/freight to a given destination are all included in the price.

Counterparty - A participant in a financial contract.

City gate - Point at which a local distribution company takes delivery of gas; physical interface between transmission and local distribution systems.

Core market - Generally that part of the gas market that does not possess fuel-switching capability in the near term; typically residential, commercial and small industrial users.

Daily balancing - (UK) Balancing, on a day-by-day basis, the amount of gas a shipper puts into a pipeline system.

Day-ahead gas - Gas for delivery on the day after the trade takes place.

Deliverability (from storage) - The rate at which gas can be supplied from a storage in a given period. In a salt cavity storage facility for example, the rate would depend on a number of factors including reservoir pressure, reservoir rock characteristics and withdrawal facilities such as pipeline capacity. The term is also used for the volume of gas which a field, pipeline, well, storage or distribution system can supply in a single 24-hour period.

Derivative - Financial instrument derived from a cash market commodity, futures contract, or other financial instrument. Derivatives can be traded on regulated exchange markets or over-the-counter. For example, energy futures contracts are derivatives of physical commodities, options on futures are derivatives of futures contracts.

DOE - US Department of Energy; Federal department.

DTI - UK Department of Trade and Industry; government ministry.

Dual-firing - Where two different fuels - e.g., gas and oil - can be used to generate energy in one piece of equipment.

Eligibility, eligible customers - Gas users that meet criteria specified in the EU Gas Directive or in national legislation, such as a minimum volume of gas consumed per year, have the right to choose their supplier and request third-party access to the grid.

EIA - Energy Information Agency; part of US DOE.

EU - European Union.

Ex-ship - Under an ex-ship contract, the seller has to deliver LNG to the buyer at an agreed importing terminal. The seller remains responsible for the LNG until it is delivered.

Exchange - Any trading arena where commodities and/or securities are bought and sold - for example, the New York Mercantile Exchange.

FERC - Federal Energy Regulatory Commission (United States); responsible for regulation of the US interstate oil and gas pipeline businesses.

Flat gas - Gas purchased with zero swing and 100% take-or-pay.

Firm capacity - Amount of gas in a buyer's contract that is guaranteed not to be interrupted.

Firm (uninterrupted) - Gas for which the full price has been paid on the understanding it will be delivered continually through the contract period.

Forward contracts - Where products are traded ahead of their physical loading.

Free-on-board (fob) - Under a fob contract, the seller provides the LNG at the exporting terminal and the buyer takes responsibility for shipping and freight insurance.

Futures contract - An exchange-traded supply contract between a buyer and a seller whereby the buyer is obligated to take delivery and the seller is obligated to provide delivery of a fixed amount of a commodity at a predetermined price at a specified location. Futures contracts are traded exclusively on regulated exchanges and are settled daily based on their current value in the marketplace.

G - Giga - 10^9.

GCV - Gross calorific value.

Gigajoule (GJ) - One billion joules, approximately equal to 948,000 British thermal units.

Gigawatt (GW) - One billion watts.

Gigawatt Hours (GWh) - One billion watt hours.

Henry Hub - The delivery point for the largest NYMEX natural gas contract by volume.

Hub - A transfer site or system where several pipelines interconnect and where shippers may obtain services to manage and facilitate their routing of supplies from production areas to markets.

IEA - International Energy Agency.

Interruptible customer - A customer that receives service only at those times and to the extent that firm customers do not demand all the available service.

Interruptible service - Gas sales that are subject to interruption for a specified number of days or hours during times of peak demand or in the event of system emergencies. In exchange for interruptibility, buyers pay lower prices.

IPE - International Petroleum Exchange, located in London.

IPP - independent power producer.

kWh - Kilowatt hour (unit of energy).

LDC (local distribution company) - (US) A company that operates or controls the retail distribution system for the delivery of natural gas or electricity.

LNG (liquefied natural gas) - Natural gas (mainly methane) which has been liquefied by reducing its temperature to minus 162 degrees Celcius at atmospheric pressure.

Load balancing - To balance demand and supply (at any given point) in a grid/pipeline/supply chain.

Load factor - The ratio of average to peak usage for gas customers for a time period i.e. one day, one hour, etc. The higher the load factor, the smaller the difference between average and peak demand.

Line-pack - Increasing the amount of gas in the system or pipeline segment by temporarily raising the pressure to meet high demand for a short period of time. Often exercised overnight as a temporary storage medium to meet anticipated next-day peaking demands.

m - mega - million - 10^6.

MBtu - Million British thermal units.

mcf - Million cubic feet.

mcf/d - Million cubic feet per day.

mcm - Million cubic metres.

mcm/d - Million cubic metres per day.

Merit order - Ranking in order of which generation plant should be used, based on ascending order of operating cost together with amount of energy that will be generated.

Mtoe - Million tonnes of oil equivalent.

MW - Megawatt.

NBP - National Balancing Point; a notional point on UK Transco's national transmission system where load is assumed to be balanced.

NEB - National Energy Board (Canada); responsible for regulation of provincial oil and gas pipelines.

NETA - New Electricity Trading Arrangements.

Netback, market value pricing - Delivered price of cheapest alternative fuel to gas to the customer (including any taxes) adjusted for any efficiency differences in the energy conversion process;

Minus cost of transporting gas from the beach/border to the customer;

Minus cost of storing gas to meeting seasonal or daily demand fluctuations;

NGTA - New Gas Trading Arrangements.

Nomination - The notification to put into effect a contract or part of a contract, e.g., a gas flow nomination from a shipper to advise the pipeline owner of the amount of gas it wishes to transport or hold in storage on a given day.

NTS - (UK) National Transmission System.

NYMEX - New York Mercantile Exchange, where futures contracts for gas and other commodities are traded.

OECD - Organisation for Economic Cooperation and Development.

Off-take - Actual amount of gas withdrawn.

OTC - Over-the-counter - An over-the-counter deal is a customised derivative contract usually arranged with an intermediary such as a major bank or the trading arm of an energy major, as opposed to a standardised derivative contract traded on an exchange. Swaps are the commonest form of OTC instrument.

Open interest - The number of futures or options natural gas contracts outstanding in the market.

One-in-twenty (1 in 20) - The highest gas demand expected on any given day over a 20 year period.

One-in-fifty (1 in 50) - The highest gas demand expected in one single year out of 50 years.

Peak day - The day during which the greatest gas demand occurs in a one year period.

Peak load - Periods during the day when energy consumption is highest. The introduction of additional gas to cover this demand is known as peak shaving.

Peak shaving - During times of peak demand, supplies from sources other than normal suppliers are used to reduce demand on the system - e.g., LNG peak shaving facilities or storage from a salt cavern.

Reserves-to-production ratio - (R/P) Remaining reserves divided by annual production.

Seasonal supplies - Supplies of gas used for winter demand. This often includes gas from storage systems.

Seasonality - All energy futures markets are affected to some extent by an annual seasonal cycle or "seasonality". This seasonal cycle or pattern refers to the tendency of market prices to move in a given direction at certain times of the year.

In the survey, "seasonality" in gas delivery is the ratio between gas consumption in the peak and the lowest month of a given year.

Shipper - A company which transports gas along a pipeline system. Shippers need to be registered with the local regulatory body.

Spark spread - The spark spread is defined as the difference, at a particular location and at a particular point in time, between the fuel cost of generating a MWh of electricity and the price of electricity.

Spot market - The spot market is the physical/cash crude, refined product, gas or electricity market. The market for immediate delivery rather than future delivery.

Spot price - The price of a security or commodity in the cash market.

Storage capacity - The amount of gas which can be stored to cover peak and seasonal demand.

Swing - Variations in gas supply or demand. A contractual commitment allowing a buyer to vary up to specified limits the amount of gas it can take at the wellhead, beach or border; the maximum daily contract quantity is usually expressed as a percentage of the annual contract quantity (100% equates to zero swing).

In the study, the swing is the maximum gas monthly delivery divided by the average monthly delivery in a given year.

Swing factor - In gas purchasing agreements the swing factor is a measure of the flexibility to vary nominations and is expressed as a ratio of peak to average supplies.

Swing producer/supplier - A company or country which changes its gas output to meet fluctuations in market demand.

T - Tera - 10^{12}.

Take-or-Pay (ToP) - In a buyer's contract take-or-pay is the obligation to pay for a specified amount of gas whether this amount is taken or not. Depending on the contract terms under-takes or over-takes may be taken as make-up or carry forward into the next contract period. When it is credited into another contract period this is called make-up gas.

tcf - trillion cubic feet.

tcm - trillion cubic meters.

Therm - equivalent to 100,000 Btu.

TJ - Terajoules.

Tolling - Under a tolling agreement a power marketer or commercial electricity customer provides the fuel, say natural gas, to produce electricity for the marketer or customer at an agreed spark spread, and receives the rights to electricity output.

TPA - Third-party access; the right or possibility for a third party to make use of the transportation or distribution services of a pipeline company to move his own gas, while paying a set or negotiated charge.

TPES - Total primary energy supply.

Unbundling - The separation of the various components of gas businesses in order to introduce greater competition to these segments of the industry.

Volatility - A measure of the variability of a market factor, most often the price of the underlying instrument. Volatility is defined mathematically as the annualised standard deviation of the natural log of the ratio of two successive prices; the actual volatility realised over a period of time (the historical volatility) can be calculated from recorded data.

Within-day gas - Gas for delivery within the day on which the trade takes place.

Working gas - The amount of gas in a storage facility above the amount needed to maintain a constant reservoir pressure (the latter is known as cushion gas).

REFERENCES

Agency for Natural Resources and Energy (2001), *Comprehensive Review of Japanese Energy Policy*, Tokyo: Ministry of Economy, Trade and Industry (METI).

British Petroleum (2001), *BP Statistical Review of World Energy*, London.

BTU Weekly, Red Bank, various issues.

Cambridge Energy Research Associates (CERA) (1998), *The Levers of Swing: Emerging Market Opportunities in Gas Storage*, Shankari Srinivasan and Michael Stoppard, Paris.

Cedigaz (2002), *Natural Gas in the World - First Estimates for 2001*, Rueil Malmaison: Institut Français du Pétrole.

Cedigaz (2001), *Natural Gas in the World: 2001 Survey*, Rueil Malmaison: Institut Français du Pétrole.

Cedigaz News Report, Rueil Malmaison, various issues.

Commission of the European Communities (CEC) (2001a), *First Report on the Implementation of the Internal Electricity and Gas Market*, Brussels: CEC.

CEC (2001b), *Completing the Internal Energy Market*, Brussels: CEC.

CEC (2001c), *Proposal for a Directive of the European Parliament and of the Council amending Directives 96/92/EC and 98/30/EC Concerning Common Rules for the Internal Market in Electricity and Natural Gas*, Brussels: CEC.

CEC (2000), *Trading Opportunities and Promotion of Transparency in the Internal Gas Market*, study prepared by Energy Markets Limited, United Kingdom; Ramboll, Denmark, Brussels: CEC.

Commission de Régulation de l'Electricité (CRE) (2002), *Rapport d'étape sur l'ouverture du marché gazier français*, Paris.

Commission de Régulation de l'Electricité et du Gaz (CREG) (2001), *Plan indicatif de l'approvisionnement en gaz naturel*, Brussels.

Crocker K. (2002), *Improving the regulatory framework for transportation and privatisation of gas pipelines: the Australian experience*, presentation by the Minister-Counsellor of the Australian Delegation to OECD at the IEA

Cross-Border Gas Trade Conference, IEA, Paris, 26-27 March 2002 (unpublished).

Department of Trade and Industry, *Energy Trends*, London: DTI, various issues.

Energy Information Administration (EIA) (2002), *Annual Energy Outlook 2002*, Washington, DC: US Department of Energy.

EIA, *Natural Gas Monthly*, Washington, DC: US DOE, various issues.

EIA (2001a), *Natural Gas Storage in the United States in 2001: a Current Assessment and Near-Term Outlook*, Washington, DC: US DOE.

EIA (2001b), *US Natural Gas Markets: Recent Trends and Prospects for the Future*, Washington, DC: US DOE.

EIA (2001c), *Impact of Interruptible Natural Gas Services on Northeast Heating Oil Demand*, Washington, DC: US DOE.

EIA (2001d), *Natural Gas Transportation - Infrastructure Issues and Operational Trends*, Washington, DC: US DOE.

EIA (2001e), U.S. *Natural Gas Markets: Mid-Term Prospects for Natural Gas Supply*, Washington, DC: US DOE.

EIA (2000a), *The Changing Structure of the Electricity Power Industry 2000: An Update*, Washington, DC: US DOE.

EIA (2000b), *The Northeast Heating Fuel Market*, Washington, DC: US DOE.

EIA (1998*)*, *The Challenges of Electric Power Restructuring for Fuel Suppliers*, Washington, DC: US DOE.

EIA (1997a), *Natural Gas 1996 - Issues and Trends*, Washington, DC: US DOE.

EIA (1997b), *1994 Manufacturing Energy Consumption Survey*, Washington, DC: US DOE.

EIA (1995), *The Value of Underground Storage in Today's Natural Gas Industry*, Washington, DC: US DOE.

ESAI (2001), *North American Natural Gas Stockwatch*, November 2001, Boston.

ESAI, *Updates Natural Gas Markets*, various issues.

Energy Sector Management Assistance Programme (ESMAP) (1993), *Long-Term Contracts - Principles and Applications*, Washington, DC: World Bank.

Eurogas (1998), *Security of Supply of Natural Gas in Western Europe*, Brussels.

EuroHub website: www.eurohubservices.com

European Gas Matters (twice-monthly), London, various issues.

FACTS (2002), *Japan Gas Update*, Gas Insights No. 3, 6 March 2002, Honolulu: Fesharaki Associates Commercial and Technical Services, Inc.

FACTS (2001a), *Japan's Gas Market: Emerging Trends in International and Domestic Markets*, Gas Alert No. 20, 23 February 2001, Honolulu: Fesharaki Associates Commercial and Technical Services, Inc.

FACTS (2001b), *Japan's Natural Gas Outlook*, Occasional Gas Report No. 9, 6 February 2001, Honolulu: Fesharaki Associates Commercial and Technical Services, Inc.

Gas Matters Today (daily), London, various issues.

Gasunie website: www.gasunie.nl

Gas Transmission Europe (GTE) (2001), *Balancing and Storage Report*, 27 June 2001, Brussels: GTE.

GTE website: www.gte.be

International Group of Liquefied Natural Gas Importers (GIIGNL) (2002), *The LNG industry - 2001*, Paris.

GIIGNL (2001), *The LNG industry - 2000*, Paris.

Huberator web site: www.huberator.com

ILEX (2001*), What influences gas prices in the UK and why have they increased through 2000?*, January 2001, ILEX Energy Consulting.

International Petroleum Exchange (IPE) website: www.ipemarkets.com

International Energy Agency (IEA) (2002a), *World Energy Outlook 2002*, Paris: Organisation for Economic Cooperation and Development.

IEA (2002b), *Natural Gas Information 2002*, Paris: OECD.

IEA (2002c), *Energy Prices and Taxes*, Paris: OECD.

IEA (2002d), *Security of Supply in Electricity Markets - Evidence and Policy Issues*, Paris: OECD.

IEA (2002e-forthcoming), *Energy Policies of IEA Countries*, Paris: OECD.

IEA (2002f), *Summary of Results of the IEA Short-Term Fuel-Switching Survey*, Note by the Secretariat, IEA/SEQ (2002) (unpublished).

IEA (2002g), *Electricity Information 2002*, Paris: OECD.

IEA (2001a), *World Energy Outlook 2001 Insights - Assessing Today's Supplies to Fuel Tomorrow's Growth*, Paris: OECD.

IEA (2001b), *Natural Gas Information 2001*, Paris: OECD.

IEA (2001c), *Energy Statistics of OECD Countries 2001*.

IEA (2000a), *World Energy Outlook 2000*, Paris: OECD.

IEA (2000b), *Regulatory Reform: European Gas*, Paris: OECD.

IEA (1998), *Natural Gas Pricing in Competitive Markets*, Paris: OECD.

IEA (1995), *The IEA Natural Gas Security Study*, Paris: OECD.

Jensen J. (2002), *Liquefied Natural Gas in North American Gas Supply*, a presentation to the Energy Modeling Forum, 24 June 2002, Washington DC (unpublished).

Jensen J. (2001a), *North American Natural Gas Price and Demand Dynamics*, a presentation to the NAPIA/PIRA Joint Annual Conference, 11 October 2001, Palm Beach, Florida (unpublished).

Jensen J. (2001b), *Comparative Energy Costs in Power Generation*, a presentation to the Centre for Global Energy Studies, at the fifth Annual Symposium "Natural gas in the Middle East Gulf in a Global Context", 24 September 2001, Bagshot, Surrey (unpublished).

Kuroyanagi N. (2001), Keynote speech delivered by the Executive Vice-President of Chubu Electric Power Co Inc at the 4th Doha Gas Conference, 14-16 March 2001, Doha (unpublished).

National Energy Board (2001), *2000 Annual Report*, Calgary: NEB.

National Energy Board (1999), *Canadian Energy - Supply and Demand to 2025*, Calgary: NEB.

Natural Gas Exchange (NGX) website: www.ngx.com.

Office of Gas and Electricity Markets (OFGEM) (2001), *Dynegy' Inc's Proposed Acquisition of BG Storage Ltd - A Consultation Document*, London: OFGEM.

OFGEM (2000), *A Review of the Development of Competition in the Gas Storage Market*, London: OFGEM.

Oil and Gas Journal (weekly), Tulsa, OK, various issues.

Petroleum Economist (2001), *Fundamentals of the Global LNG Industry*, Petroleum Economist Ltd in association with Shell Gas & Power, London.

Petrostratregies (weekly), Paris, various issues.

Roe D. (2001), *LNG Trade: A Review of Markets, Projects and Issues in the Changing World of LNG*, SMi, Business to Business Publishing, London.

The Utilities Journal (2002), *NETA: The First Year*, March 2002, p. 30-31, Oxford.

Valais M., TotalFinaElf, Chabrelie M.F., Cedigaz, Lefeuvre T., Gaz de France (2001), *World LNG Prospects: Favourable Parameters for a New Growth Area*, presentation to the 18th World Energy Congress, 21-25 October 2001, Buenos Aires.

Van Nieuwland A.J.F.M (2001), *Planning and Developing NAM's UGS Strategy*, Presentation to the Global Gas Village Conference: The Case for Underground Gas Storage in Europe, 17-18 May 2001, Paris (unpublished).

World Gas Intelligence (weekly), New York, various issues.

ORDER FORM

IEA BOOKS

Fax: +33 (0)1 40 57 65 59
E-mail: books@iea.org
www.iea.org/books

INTERNATIONAL ENERGY AGENCY

9, rue de la Fédération
F-75739 Paris Cedex 15

I would like to order the following publications

PUBLICATIONS	ISBN	QTY	PRICE		TOTAL
☐ **Flexibility in Natural Gas Supply and Demand**	**92-64-19938-1**		**$75**	**€82**	
☐ World Energy Outlook 2002	92-64-19835-0		$150	€165	
☐ Distributed Generation in Liberalised Electricity Markets	92-64-19802-4		$75	€82	
☐ Security of Supply in Electricity Markets - *Evidence and Policy Issues*	92-64-19805-9		$100	€110	
☐ Russia Energy Survey 2002	92-64-18732-4		$150	€165	
☐ Natural Gas Information 2002	92-64-19791-5		$150	€165	
☐ Developing China's Natural Gas Market - *The Energy Policy Challenges*	92-64-19837-7		$100	€110	
☐ South American Gas - *Daring to Tap the Bounty* (Forthcoming)	92-64-19663-3		$120	€132	
			TOTAL		

DELIVERY DETAILS

Name _____   Organisation _____

Address _____

Country _____   Postcode _____

Telephone _____   E-mail _____

PAYMENT DETAILS

☐ I enclose a cheque payable to IEA Publications for the sum of $ _____ or € _____

☐ Please debit my credit card (tick choice).   ☐ Mastercard   ☐ VISA   ☐ American Express

Card no: ⌊_⌊_⌊_⌊_⌊_⌊_⌊_⌊_⌊_⌊_⌊_⌊_⌊_⌋

Expiry date: ⌊_⌊_⌊_⌊_⌋   Signature: _____

OECD PARIS CENTRE
Tel: (+33-01) 45 24 81 67
Fax: (+33-01) 49 10 42 76
E-mail: distribution@oecd.org

OECD BONN CENTRE
Tel: (+49-228) 959 12 15
Fax: (+49-228) 959 12 18
E-mail: bonn.contact@oecd.org

OECD MEXICO CENTRE
Tel: (+52-5) 280 12 09
Fax: (+52-5) 280 04 80
E-mail: mexico.contact@oecd.org

You can also send your order to your nearest OECD sales point or through the OECD online services:
www.oecd.org/ bookshop

OECD TOKYO CENTRE
Tel: (+81-3) 3586 2016
Fax: (+81-3) 3584 7929
E-mail: center@oecdtokyo.org

OECD WASHINGTON CENTER
Tel: (+1-202) 785-6323
Toll-free number for orders:
(+1-800) 456-6323
Fax: (+1-202) 785-0350
E-mail: washington.contact@oecd.org

IEA PUBLICATIONS, 9, Rue de la Fédération, 75739 PARIS cedex 15
PRINTED IN FRANCE BY JOUVE
(61 2002 321 PI) ISBN : 92-64-19938-1 – 2002

Book Cover by Bertrand Sadin